Student Resources

Harcourt

INCLUDES
- Program Authors
- Table of Contents
- Glossary
- Common Core State Standards Correlation
- Index
- Table of Measures

Made in the United States
Text printed on 100% recycled paper

Houghton Mifflin Harcourt

GO MATH!

Copyright © by Houghton Mifflin Harcourt Publishing Company

All rights reserved. No part of this work may be reproduced or transmitted in any form or by any means, electronic or mechanical, including photocopying or recording, or by any information storage or retrieval system, without the prior written permission of the copyright owner unless such copying is expressly permitted by federal copyright law.

Permission is hereby granted to individuals using the corresponding student's textbook or kit as the major vehicle for regular classroom instruction to photocopy entire pages from this publication in classroom quantities for instructional use and not for resale. Requests for information on other matters regarding duplication of this work should be addressed to Houghton Mifflin Harcourt Publishing Company, Attn: Contracts, Copyrights, and Licensing, 9400 Southpark Center Loop, Orlando, Florida 32819-8647.

Common Core State Standards © Copyright 2010. National Governors Association Center for Best Practices and Council of Chief State School Officers. All rights reserved.

This product is not sponsored or endorsed by the Common Core State Standards Initiative of the National Governors Association Center for Best Practices and the Council of Chief State School Officers.

Printed in the U.S.A.

ISBN 978-0-544-34348-1

26 0858 22

4500855735   B C D E F G

If you have received these materials as examination copies free of charge, Houghton Mifflin Harcourt Publishing Company retains title to the materials and they may not be resold. Resale of examination copies is strictly prohibited.

Possession of this publication in print format does not entitle users to convert this publication, or any portion of it, into electronic format.

Dear Students and Families,

Welcome to **Go Math!**, Grade 5! In this exciting mathematics program, there are hands-on activities to do and real-world problems to solve. Best of all, you will write your ideas and answers right in your book. In **Go Math!**, writing and drawing on the pages helps you think deeply about what you are learning, and you will really understand math!

By the way, all of the pages in your **Go Math!** book are made using recycled paper. We wanted you to know that you can Go Green with **Go Math!**

Sincerely,

The Authors

Made in the United States
Text printed on 100% recycled paper

GO MATH!

Authors

Juli K. Dixon, Ph.D.
Professor, Mathematics Education
University of Central Florida
Orlando, Florida

Edward B. Burger, Ph.D.
President, Southwestern University
Georgetown, Texas

Steven J. Leinwand
Principal Research Analyst
American Institutes for
  Research (AIR)
Washington, D.C.

Contributor

Rena Petrello
Professor, Mathematics
Moorpark College
Moorpark, CA

Matthew R. Larson, Ph.D.
K-12 Curriculum Specialist for
  Mathematics
Lincoln Public Schools
Lincoln, Nebraska

Martha E. Sandoval-Martinez
Math Instructor
El Camino College
Torrance, California

English Language Learners Consultant

Elizabeth Jiménez
CEO, GEMAS Consulting
Professional Expert on English
  Learner Education
Bilingual Education and
  Dual Language
Pomona, California

Table of Contents

Student Edition Table of Contents.................... v

Glossary.. H1

Common Core State Standards Correlation H14

Index.. H24

Table of Measures.................................. H37

Fluency with Whole Numbers and Decimals

 Critical Area Extending division to 2-digit divisors, integrating decimal fractions into the place value system and developing understanding of operations with decimals to hundredths, and developing fluency with whole number and decimal operations

 In the Chef's Kitchen 2

1 Place Value, Multiplication, and Expressions 3

Domains Operations and Algebraic Thinking
Number and Operations in Base Ten
COMMON CORE STATE STANDARDS
5.OA.A.1, 5.OA.A.2, 5.NBT.A.1, 5.NBT.A.2, 5.NBT.B.5, 5.NBT.B.6

✓ Show What You Know 3
  Vocabulary Builder .4
  Chapter Vocabulary Cards
  Vocabulary Game 4A
1 **Investigate** • Place Value and Patterns 5
2 Place Value of Whole Numbers 11
3 **Algebra** • Properties. 17
4 **Algebra** • Powers of 10 and Exponents 23
5 **Algebra** • Multiplication Patterns 29
✓ Mid-Chapter Checkpoint 35
6 Multiply by 1-Digit Numbers 37
7 Multiply by Multi-Digit Numbers 43
8 Relate Multiplication to Division 49
9 **Problem Solving** • Multiplication and Division 55
10 **Algebra** • Numerical Expressions 61
11 **Algebra** • Evaluate Numerical Expressions. 67
12 **Algebra** • Grouping Symbols 73
✓ Chapter 1 Review/Test. 79

GO DIGITAL
Go online! Your math lessons are interactive. Use iTools, Animated Math Models, the Multimedia eGlossary, and more.

Chapter 1 Overview
In this chapter, you will explore and discover answers to the following **Essential Questions**:

• How can you use place value, multiplication, and expressions to represent and solve problems?
• How can you read, write, and represent whole numbers through millions?
• How can you use properties and multiplication to solve problems?
• How can you use expressions to represent and solve a problem?

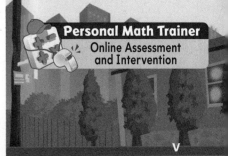

Chapter 2 Overview

In this chapter, you will explore and discover answers to the following **Essential Questions**:

- How can you divide whole numbers?
- What strategies have you used to place the first digit in the quotient?
- How can you use estimation to help you divide?
- How do you know when to use division to solve a problem?

Practice and Homework

Lesson Check and Spiral Review in every lesson

Chapter 3 Overview

In this chapter, you will explore and discover answers to the following **Essential Questions**:

- How can you add and subtract decimals?
- What methods can you use to find decimal sums and differences?
- How does using place value help you add and subtract decimals?

2 Divide Whole Numbers — 85

Domains Number and Operations in Base Ten
Number and Operations–Fractions
COMMON CORE STATE STANDARDS
5.NBT.B.6, 5.NF.B.3

✓ Show What You Know 85
Vocabulary Builder 86
Chapter Vocabulary Cards
Vocabulary Game 86A
1 Place the First Digit 87
2 Divide by 1-Digit Divisors 93
3 Investigate • Division with 2-Digit Divisors 99
4 Partial Quotients 105
✓ Mid-Chapter Checkpoint 111
5 Estimate with 2-Digit Divisors. 113
6 Divide by 2-Digit Divisors 119
7 Interpret the Remainder 125
8 Adjust Quotients 131
9 Problem Solving • Division 137
✓ Chapter 2 Review/Test 143

3 Add and Subtract Decimals — 149

Domain Number and Operations in Base Ten
COMMON CORE STATE STANDARDS
5.NBT.A.1, 5.NBT.A.3a, 5.NBT.A.3b, 5.NBT.A.4, 5.NBT.B.7

✓ Show What You Know 149
Vocabulary Builder 150
Chapter Vocabulary Cards
Vocabulary Game 150A
1 Investigate • Thousandths 151
2 Place Value of Decimals 157
3 Compare and Order Decimals 163
4 Round Decimals 169
5 Investigate • Decimal Addition 175
6 Investigate • Decimal Subtraction 181
✓ Mid-Chapter Checkpoint 187
7 Estimate Decimal Sums and Differences 189
8 Add Decimals 195
9 Subtract Decimals 201
10 Algebra • Patterns with Decimals 207
11 Problem Solving • Add and Subtract Money 213
12 Choose a Method 219
✓ Chapter 3 Review/Test 225

4 Multiply Decimals — 231

Domain Number and Operations in Base Ten
COMMON CORE STATE STANDARDS
5.NBT.A.2, 5.NBT.B.7

- ✓ Show What You Know 231
- Vocabulary Builder 232
- Chapter Vocabulary Cards
- Vocabulary Game 232A
- **1 Algebra** • Multiplication Patterns with Decimals 233
- **2 Investigate** • Multiply Decimals and Whole Numbers 239
- **3** Multiplication with Decimals and Whole Numbers 245
- **4** Multiply Using Expanded Form 251
- **5 Problem Solving** • Multiply Money 257
- ✓ Mid-Chapter Checkpoint 263
- **6 Investigate** • Decimal Multiplication 265
- **7** Multiply Decimals 271
- **8** Zeros in the Product 277
- ✓ Chapter 4 Review/Test 283

Chapter 4 Overview

In this chapter, you will explore and discover answers to the following **Essential Questions**:

- How can you solve decimal multiplication problems?
- How is multiplying with decimals similar to multiplying with whole numbers?
- How can patterns, models, and drawings help you solve decimal multiplication problems?
- How do you know where to place a decimal point in a product?
- How do you know the correct number of decimal places in a product?

5 Divide Decimals — 289

Domain Number and Operations in Base Ten
COMMON CORE STATE STANDARDS
5.NBT.A.2, 5.NBT.B.7

- ✓ Show What You Know 289
- Vocabulary Builder 290
- Chapter Vocabulary Cards
- Vocabulary Game 290A
- **1 Algebra** • Division Patterns with Decimals 291
- **2 Investigate** • Divide Decimals by Whole Numbers 297
- **3** Estimate Quotients 303
- **4** Division of Decimals by Whole Numbers 309
- ✓ Mid-Chapter Checkpoint 315
- **5 Investigate** • Decimal Division 317
- **6** Divide Decimals 323
- **7** Write Zeros in the Dividend 329
- **8 Problem Solving** • Decimal Operations 335
- ✓ Chapter 5 Review/Test 341

Chapter 5 Overview

In this chapter, you will explore and discover answers to the following **Essential Questions**:

- How can you solve decimal division problems?
- How is dividing with decimals similar to dividing with whole numbers?
- How can patterns, models, and drawings help you solve decimal division problems?
- How do you know where to place a decimal point in a quotient?
- How do you know the correct number of decimal places in a quotient?

Critical Area: Operations with Fractions

Critical Area Developing fluency with addition and subtraction of fractions, and developing understanding of the multiplication of fractions and of division of fractions in limited cases (unit fractions divided by whole numbers and whole numbers divided by unit fractions)

 The Rhythm Track 348

6 Add and Subtract Fractions with Unlike Denominators — 349

Domains Operations and Algebraic Thinking
Number and Operations–Fractions

COMMON CORE STATE STANDARDS
5.OA.A.2, 5.NF.A.1, 5.NF.A.2

- ✓ Show What You Know 349
- Vocabulary Builder 350
- Chapter Vocabulary Cards
- Vocabulary Game 350A
- **1** Investigate • Addition with Unlike Denominators 351
- **2** Investigate • Subtraction with Unlike Denominators 357
- **3** Estimate Fraction Sums and Differences 363
- **4** Common Denominators and Equivalent Fractions 369
- **5** Add and Subtract Fractions 375
- ✓ Mid-Chapter Checkpoint 381
- **6** Add and Subtract Mixed Numbers 383
- **7** Subtraction with Renaming 389
- **8** Algebra • Patterns with Fractions 395
- **9** Problem Solving • Practice Addition and Subtraction 401
- **10** Algebra • Use Properties of Addition 407
- ✓ Chapter 6 Review/Test 413

GO DIGITAL
Go online! Your math lessons are interactive. Use iTools, Animated Math Models, the Multimedia eGlossary, and more.

Chapter 6 Overview

In this chapter, you will explore and discover answers to the following **Essential Questions**:

- How can you add and subtract fractions with unlike denominators?
- How do models help you find sums and differences of fractions?
- When you add and subtract fractions, when do you use the least common denominator?

Personal Math Trainer
Online Assessment and Intervention

7 Multiply Fractions — 419

Domain Number and Operations–Fractions
COMMON CORE STATE STANDARDS
5.NF.B.4a, 5.NF.B.4b, 5.NF.B.5a, 5.NF.B.5b, 5.NF.B.6

- ✓ Show What You Know 419
- Vocabulary Builder 420
- Chapter Vocabulary Cards
- Vocabulary Game 420A
- 1 Find Part of a Group 421
- 2 *Investigate* • Multiply Fractions and Whole Numbers 427
- 3 Fraction and Whole Number Multiplication 433
- 4 *Investigate* • Multiply Fractions 439
- 5 Compare Fraction Factors and Products 445
- 6 Fraction Multiplication 451
- ✓ Mid-Chapter Checkpoint 457
- 7 *Investigate* • Area and Mixed Numbers 459
- 8 Compare Mixed Number Factors and Products 465
- 9 Multiply Mixed Numbers 471
- 10 Problem Solving • Find Unknown Lengths 477
- ✓ Chapter 7 Review/Test 483

Chapter 7 Overview
In this chapter, you will explore and discover answers to the following **Essential Questions**:
- How do you multiply fractions?
- How can you model fraction multiplication?
- How can you compare fraction factors and products?

Practice and Homework
Lesson Check and Spiral Review in every lesson

8 Divide Fractions — 489

Domain Number and Operations–Fractions
COMMON CORE STATE STANDARDS
5.NF.B.3, 5.NF.B.7a, 5.NF.B.7b, 5.NF.B.7c

- ✓ Show What You Know 489
- Vocabulary Builder 490
- Chapter Vocabulary Cards
- Vocabulary Game 490A
- 1 *Investigate* • Divide Fractions and Whole Numbers 491
- 2 Problem Solving • Use Multiplication 497
- 3 Connect Fractions to Division 503
- ✓ Mid-Chapter Checkpoint 509
- 4 Fraction and Whole-Number Division 511
- 5 Interpret Division with Fractions 517
- ✓ Chapter 8 Review/Test 523

Chapter 8 Overview
In this chapter, you will explore and discover answers to the following **Essential Questions**:
- What strategies can you use to solve division problems involving fractions?
- What is the relationship between multiplication and division, and how can you use it to solve division problems?
- How can you use fractions, diagrams, equations, and story problems to represent division?
- When you divide a whole number by a fraction or a fraction by a whole number, how do the dividend, the divisor, and the quotient compare?

Critical Area

Geometry and Measurement

 Critical Area Developing understanding of volume

GO DIGITAL

Go online! Your math lessons are interactive. Use *i*Tools, Animated Math Models, the Multimedia eGlossary, and more.

Chapter 9 Overview

In this chapter, you will explore and discover answers to the following **Essential Questions**:

- How can you use line plots, coordinate grids, and patterns to help you graph and interpret data?
- How can a line plot help you find an average with data given in fractions?
- How can a coordinate grid help you interpret experimental and real-world data?
- How can you write and graph ordered pairs on a coordinate grid using two numerical patterns?

Personal Math Trainer
Online Assessment and Intervention

Project Space Architecture 530

9 Algebra: Patterns and Graphing   531

Domains Operations and Algebraic Thinking
Measurement and Data
Geometry

COMMON CORE STATE STANDARDS
5.OA.B.3, 5.MD.B.2, 5.G.A.1, 5.G.A.2

✓ Show What You Know 531
Vocabulary Builder 532
Chapter Vocabulary Cards
Vocabulary Game 532A
1 Line Plots 533
2 Ordered Pairs 539
3 Investigate • Graph Data 545
4 Line Graphs 551
✓ Mid-Chapter Checkpoint 557
5 Numerical Patterns 559
6 Problem Solving • Find a Rule 565
7 Graph and Analyze Relationships 571
✓ Chapter 9 Review/Test 577

10 Convert Units of Measure — 583

Domain Measurement and Data
COMMON CORE STATE STANDARD
5.MD.A.1

- ✓ Show What You Know 583
  - Vocabulary Builder 584
  - Chapter Vocabulary Cards
  - Vocabulary Game 584A
- 1 Customary Length 585
- 2 Customary Capacity 591
- 3 Weight 597
- 4 Multistep Measurement Problems 603
- ✓ Mid-Chapter Checkpoint 609
- 5 Metric Measures 611
- 6 Problem Solving • Customary and Metric Conversions 617
- 7 Elapsed Time 623
- ✓ Chapter 10 Review/Test 629

Chapter 10 Overview

In this chapter, you will explore and discover answers to the following **Essential Questions**:

- What strategies can you use to compare and convert measurements?
- How can you decide whether to multiply or divide when you are converting measurements?
- How can you organize your solution when you are solving a multistep measurement problem?
- How is converting metric measurements different from converting customary measurements?

Practice and Homework

Lesson Check and Spiral Review in every lesson

Chapter 11 Overview

In this chapter, you will explore and discover answers to the following **Essential Questions**:

- How do unit cubes help you build solid figures and understand the volume of a rectangular prism?
- How can you identify, describe, and classify three-dimensional figures?
- How can you find the volume of a rectangular prism?

Geometry and Volume — 635

Domains Measurement and Data
Geometry

COMMON CORE STATE STANDARDS
5.MD.C.3, 5.MD.C.3a, 5.MD.C.3b, 5.MD.C.4, 5.MD.C.5a, 5.MD.C.5b, 5.MD.C.5c, 5.G.B.3, 5.G.B.4

✓ Show What You Know 635
Vocabulary Builder 636
Chapter Vocabulary Cards
Vocabulary Game 636A
1 Polygons 637
2 Triangles 643
3 Quadrilaterals 649
4 Three-Dimensional Figures 655
✓ Mid-Chapter Checkpoint 661
5 Investigate • Unit Cubes and Solid Figures 663
6 Investigate • Understand Volume 669
7 Investigate • Estimate Volume 675
8 Volume of Rectangular Prisms 681
9 Algebra • Apply Volume Formulas 687
10 Problem Solving • Compare Volumes 693
11 Find Volume of Composed Figures 699
✓ Chapter 11 Review/Test 705

Glossary H1
Common Core State Standards for
Mathematics Correlations H14
Index H24
Table of Measures H37

Glossary

Pronunciation Key

a	add, map	ē	equal, tree	m	move, seem	oo	pool, food	u̇	pull, book
ā	ace, rate	f	fit, half	n	nice, tin	p	pit, stop	û(r)	burn, term
â(r)	care, air	g	go, log	ng	ring, song	r	run, poor	yoo	fuse, few
ä	palm, father	h	hope, hate	o	odd, hot	s	see, pass	v	vain, eve
b	bat, rub	i	it, give	ō	open, so	sh	sure, rush	w	win, away
ch	check, catch	ī	ice, write	ô	order, jaw	t	talk, sit	y	yet, yearn
d	dog, rod	j	joy, ledge	oi	oil, boy	th	thin, both	z	zest, muse
e	end, pet	k	cool, take	ou	pout, now	<u>th</u>	this, bathe	zh	vision, pleasure
		l	look, rule	o͝o	took, full	u	up, done		

ə the schwa, an unstressed vowel representing the sound spelled a in **a**bove, e in sick**e**n, i in poss**i**ble, o in mel**o**n, u in circ**u**s

Other symbols:
• separates words into syllables
′ indicates stress on a syllable

acute angle [ə•kyoot′ ang′gəl] **ángulo agudo** An angle that has a measure less than a right angle (less than 90° and greater than 0°)
Example:

Word History

The Latin word for needle is *acus*. This means "pointed" or "sharp." You will recognize the root in the words *acid* (sharp taste), *acumen* (mental sharpness), and *acute*, which describes a sharp or pointed angle.

acute triangle [ə•kyoot′ trī′ang•gəl] **triángulo acutángulo** A triangle that has three acute angles

addend [ad′end] **sumando** A number that is added to another in an addition problem

addition [ə•dish′ən] **suma** The process of finding the total number of items when two or more groups of items are joined; the inverse operation of subtraction

algebraic expression [al•jə•brā′ik ek•spresh′ən] **expresión algebraica** An expression that includes at least one variable
Examples: $x + 5$, $3a - 4$

angle [ang′gəl] **ángulo** A shape formed by two rays that share the same endpoint
Example:

area [âr′ē•ə] **área** The measure of the number of unit squares needed to cover a surface

array [ə•rā′] **matriz** An arrangement of objects in rows and columns
Example:

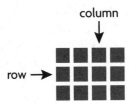

Student Handbook H1

Associative Property of Addition [ə•sō'shē•āt•iv präp'ər•tē əv ə•dish'ən] **propiedad asociativa de la suma** The property that states that when the grouping of addends is changed, the sum is the same
Example: (5 + 8) + 4 = 5 + (8 + 4)

Associative Property of Multiplication [ə•sō'shē•āt•iv präp'ər•tē əv mul•tə•pli•kā'shən] **propiedad asociativa de la multiplicación** The property that states that factors can be grouped in different ways and still get the same product
Example: (2 × 3) × 4 = 2 × (3 × 4)

balance [bal'əns] **equilibrar** To equalize in weight or number

bar graph [bär graf] **gráfica de barras** A graph that uses horizontal or vertical bars to display countable data
Example:

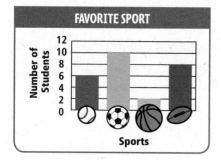

base (arithmetic) [bās] **base** A number used as a repeated factor
Example: $8^3 = 8 \times 8 \times 8$. The base is 8.

base (geometry) [bās] **base** In two dimensions, one side of a triangle or parallelogram that is used to help find the area. In three dimensions, a plane figure, usually a polygon or circle, by which a three-dimensional figure is measured or named
Examples:

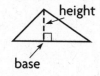

benchmark [bench'märk] **punto de referencia** A familiar number used as a point of reference

capacity [kə•pas'i•tē] **capacidad** The amount a container can hold when filled

Celsius (°C) [sel'sē•əs] **Celsius (°C)** A metric scale for measuring temperature

centimeter (cm) [sen'tə•mēt•ər] **centímetro (cm)** A metric unit used to measure length or distance; 0.01 meter = 1 centimeter

closed figure [klōzd fig'yər] **figura cerrada** A figure that begins and ends at the same point

common denominator [käm'ən dē•näm'ə•nāt•ər] **denominador común** A common multiple of two or more denominators
Example: Some common denominators for $\frac{1}{4}$ and $\frac{5}{6}$ are 12, 24, and 36.

common factor [käm'ən fak'tər] **factor común** A number that is a factor of two or more numbers

common multiple [käm'ən mul'tə•pəl] **múltiplo común** A number that is a multiple of two or more numbers

Commutative Property of Addition [kə•myōōt'ə•tiv präp'ər•tē əv ə•dish'ən] **propiedad conmutativa de la suma** The property that states that when the order of two addends is changed, the sum is the same
Example: 4 + 5 = 5 + 4

Commutative Property of Multiplication [kə•myōōt'ə•tiv präp'ər•tē əv mul•tə•pli•kā'shən] **propiedad conmutativa de la multiplicación** The property that states that when the order of two factors is changed, the product is the same
Example: 4 × 5 = 5 × 4

compatible numbers [kəm•pat'ə•bəl num'bərz] **números compatibles** Numbers that are easy to compute with mentally

composite number [kəm•päz'it num'bər] **número compuesto** A number having more than two factors
Example: 6 is a composite number, since its factors are 1, 2, 3, and 6.

cone [kōn] **cono** A solid figure that has a flat, circular base and one vertex
Example:

congruent [kən·grōō′ənt] **congruente** Having the same size and shape

coordinate grid [kō·ôrd′n·it grid] **cuadrícula de coordenadas** A grid formed by a horizontal line called the *x*-axis and a vertical line called the *y*-axis
Example:

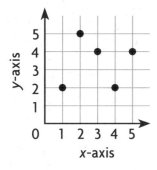

counting number [kount′ing num′bər] **número natural** A whole number that can be used to count a set of objects (1, 2, 3, 4, . . .)

cube [kyōōb] **cubo** A three-dimensional figure with six congruent square faces
Example:

cubic unit [kyōō′bik yōō′nit] **unidad cúbica** A unit used to measure volume such as cubic foot (ft³), cubic meter (m³), and so on

cup (c) [kup] **taza (t)** A customary unit used to measure capacity; 8 ounces = 1 cup

cylinder [sil′ən·dər] **cilindro** A solid figure that has two parallel bases that are congruent circles
Example:

data [dāt′ə] **datos** Information collected about people or things, often to draw conclusions about them

decagon [dek′ə·gän] **decágono** A polygon with ten sides and ten angles
Examples:

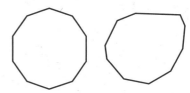

decagonal prism [dek·ag′ə·nəl priz′əm] **prisma decagonal** A three-dimensional figure with two decagonal bases and ten rectangular faces

decimal [des′ə·məl] **decimal** A number with one or more digits to the right of the decimal point

decimal point [des′ə·məl point] **punto decimal** A symbol used to separate dollars from cents in money, and to separate the ones place from the tenths place in a decimal

decimal system [des′ə·məl sis′təm] **sistema decimal** A system of computation based on the number 10

decimeter (dm) [des′i·mēt·ər] **decímetro (dm)** A metric unit used to measure length or distance; 10 decimeters = 1 meter

degree (°) [di·grē′] **grado (°)** A unit used for measuring angles and temperature

degree Celsius (°C) [di·grē′ sel′sē·əs] **grado Celsius** A metric unit for measuring temperature

degree Fahrenheit (°F) [di·grē′ fâr′ən·hīt] **grado Fahrenheit** A customary unit for measuring temperature

dekameter (dam) [dek′ə·mēt·ər] **decámetro** A metric unit used to measure length or distance; 10 meters = 1 dekameter

denominator [dē·näm′ə·nāt·ər] **denominador** The number below the bar in a fraction that tells how many equal parts are in the whole or in the group
Example: $\frac{3}{4}$ ← denominator

Student Handbook H3

diagonal [dī·ag′ə·nəl] **diagonal** A line segment that connects two non-adjacent vertices of a polygon
Example:

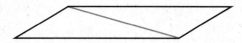

difference [dif′ər·əns] **diferencia** The answer to a subtraction problem

digit [dij′it] **dígito** Any one of the ten symbols 0, 1, 2, 3, 4, 5, 6, 7, 8, 9 used to write numbers

dimension [də·men′shən] **dimensión** A measure in one direction

Distributive Property [di·strib′yōō·tiv präp′ər·tē] **propiedad distributiva** The property that states that multiplying a sum by a number is the same as multiplying each addend in the sum by the number and then adding the products
Example: $3 \times (4 + 2) = (3 \times 4) + (3 \times 2)$
$3 \times 6 = 12 + 6$
$18 = 18$

divide [də·vīd′] **dividir** To separate into equal groups; the inverse operation of multiplication

dividend [div′ə·dend] **dividendo** The number that is to be divided in a division problem
Example: $36 \div 6$; $6\overline{)36}$ The dividend is 36.

division [də·vizh′ən] **división** The process of sharing a number of items to find how many equal groups can be made or how many items will be in each equal group; the inverse operation of multiplication

divisor [də·vī′zər] **divisor** The number that divides the dividend
Example: $15 \div 3$; $3\overline{)15}$ The divisor is 3.

edge [ej] **arista** The line segment made where two faces of a solid figure meet
Example:

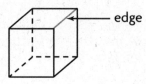

elapsed time [ē·lapst′ tīm] **tiempo transcurrido** The time that passes between the start of an activity and the end of that activity

endpoint [end′ point] **extremo** The point at either end of a line segment or the starting point of a ray

equal to (=) [ē′kwəl tōō] **igual a** Having the same value

equation [ē·kwā′zhən] **ecuación** An algebraic or numerical sentence that shows that two quantities are equal

equilateral triangle [ē·kwi·lat′ər·əl trī′ang·gəl] **triángulo equilátero** A triangle with three congruent sides
Example:

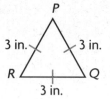

equivalent [ē·kwiv′ə·lənt] **equivalente** Having the same value

equivalent decimals [ē·kwiv′ə·lənt des′ə·məlz] **decimales equivalentes** Decimals that name the same amount
Example: $0.4 = 0.40 = 0.400$

equivalent fractions [ē·kwiv′ə·lənt frak′shənz] **fracciones equivalentes** Fractions that name the same amount or part
Example: $\frac{3}{4} = \frac{6}{8}$

estimate [es′tə·mit] *noun* **estimación (s)** A number close to an exact amount

estimate [es′tə·māt] *verb* **estimar (v)** To find a number that is close to an exact amount

evaluate [ē·val′yōō·āt] **evaluar** To find the value of a numerical or algebraic expression

even [ē′vən] **par** A whole number that has a 0, 2, 4, 6, or 8 in the ones place

expanded form [ek·span′did fôrm] **forma desarrollada** A way to write numbers by showing the value of each digit
Examples: $832 = 8 \times 100 + 3 \times 10 + 2 \times 1$
$3.25 = (3 \times 1) + (2 \times \frac{1}{10}) + (5 \times \frac{1}{100})$

exponent [eks′·pōn·ənt] **exponente** A number that shows how many times the base is used as a factor
Example: $10^3 = 10 \times 10 \times 10$. 3 is the exponent.

expression [ek·spresh′ən] **expresión** A mathematical phrase or the part of a number sentence that combines numbers, operation signs, and sometimes variables, but does not have an equal sign

face [fās] **cara** A polygon that is a flat surface of a solid figure
Example:

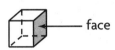

fact family [fakt fam′ə·lē] **familia de operaciones** A set of related multiplication and division, or addition and subtraction, equations
Examples: $7 \times 8 = 56; 8 \times 7 = 56;$
$56 \div 7 = 8; 56 \div 8 = 7$

factor [fak′tər] **factor** A number multiplied by another number to find a product

Fahrenheit (°F) [fâr′ən·hīt] **Fahrenheit (°F)** A customary scale for measuring temperature

fluid ounce (fl oz) [floo′id ouns] **onza fluida** A customary unit used to measure liquid capacity; 1 cup = 8 fluid ounces

foot (ft) [foot] **pie (ft)** A customary unit used to measure length or distance; 1 foot = 12 inches

formula [fôr′myoo·lə] **fórmula** A set of symbols that expresses a mathematical rule
Example: $A = b \times h$

fraction [frak′shən] **fracción** A number that names a part of a whole or a part of a group

fraction greater than 1 [frak′shən grāt′ər than wun] **fracción mayor que 1** A number which has a numerator that is greater than its denominator
Example:

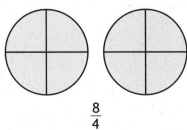

$\frac{8}{4}$

gallon (gal) [gal′ən] **galón (gal)** A customary unit used to measure capacity; 4 quarts = 1 gallon

general quadrilateral [jen′ər·əl kwä·dri·lat′ər·əl] **cuadrilátero en general** See *quadrilateral*.

gram (g) [gram] **gramo (g)** A metric unit used to measure mass; 1,000 grams = 1 kilogram

greater than (>) [grāt′ər than] **mayor que (>)** A symbol used to compare two numbers or two quantities when the greater number or greater quantity is given first
Example: $6 > 4$

greater than or equal to (≥) [grāt′ər than ôr ē′kwəl too] **mayor que o igual a** A symbol used to compare two numbers or quantities when the first is greater than or equal to the second

greatest common factor [grāt′əst käm′ən fak′tər] **máximo común divisor** The greatest factor that two or more numbers have in common
Example: 6 is the greatest common factor of 18 and 30.

grid [grid] **cuadrícula** Evenly divided and equally spaced squares on a figure or flat surface

height [hīt] **altura** The length of a perpendicular from the base to the top of a two-dimensional or three-dimensional figure
Example:

heptagon [hep′tə·gän] **heptágono** A polygon with seven sides and seven angles

Student Handbook H5

hexagon [hek′sə•gän] **hexágono** A polygon with six sides and six angles
Examples:

hexagonal prism [hek•sag′ə•nəl priz′əm] **prisma hexagonal** A three-dimensional figure with two hexagonal bases and six rectangular faces

horizontal [hôr•i•zänt′l] **horizontal** Extending left and right

hundredth [hun′drədth] **centésimo** One of 100 equal parts
Examples: 0.56, $\frac{56}{100}$, fifty-six hundredths

Identity Property of Addition [ī•den′tə•tē präp′ər•tē əv ə•dish′ən] **propiedad de identidad de la suma** The property that states that when you add zero to a number, the result is that number

Identity Property of Multiplication [ī•den′tə•tē präp′ər•tē əv mul•tə•pli•kā′shən] **propiedad de identidad de la multiplicación** The property that states that the product of any number and 1 is that number

inch (in.) [inch] **pulgada (pulg)** A customary unit used to measure length or distance; 12 inches = 1 foot

inequality [in•ē•kwôl′ə•tē] **desigualdad** A mathematical sentence that contains the symbol <, >, ≤, ≥, or ≠

intersecting lines [in•tər•sekt′ing līnz] **líneas secantes** Lines that cross each other at exactly one point
Example:

interval [in′tər•vəl] **intervalo** The difference between one number and the next on the scale of a graph

inverse operations [in′vûrs äp•ə•rā′shənz] **operaciones inversas** Opposite operations, or operations that undo each other, such as addition and subtraction or multiplication and division

isosceles triangle [ī•säs′ə•lēz trī′ang•gəl] **triángulo isósceles** A triangle with two congruent sides
Example:

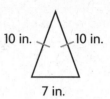

key [kē] **clave** The part of a map or graph that explains the symbols

kilogram (kg) [kil′ō•gram] **kilogramo (kg)** A metric unit used to measure mass; 1,000 grams = 1 kilogram

kilometer (km) [kə•läm′ət•ər] **kilómetro (km)** A metric unit used to measure length or distance; 1,000 meters = 1 kilometer

lateral face [lat′ər•əl fās] **cara lateral** Any surface of a polyhedron other than a base

least common denominator [lēst käm′ən dē•näm′ə•nāt•ər] **mínimo común denominador** The least common multiple of two or more denominators
Example: The least common denominator for $\frac{1}{4}$ and $\frac{5}{6}$ is 12.

least common multiple [lēst käm′ən mul′tə•pəl] **mínimo común múltiplo** The least number that is a common multiple of two or more numbers

less than (<) [les <u>th</u>an] **menor que (<)** A symbol used to compare two numbers or two quantities, with the lesser number given first
Example: 4 < 6

less than or equal to (≤) [les than ôr ēʹkwəl too] **menor que o igual a** A symbol used to compare two numbers or two quantities, when the first is less than or equal to the second

line [līn] **línea** A straight path in a plane, extending in both directions with no endpoints
Example:

⟵――――――――⟶

line graph [līn graf] **gráfica lineal** A graph that uses line segments to show how data change over time

line plot [līn plät] **diagrama de puntos** A graph that shows frequency of data along a number line
Example:

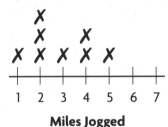

line segment [līn segʹmənt] **segmento** A part of a line that includes two points called endpoints and all the points between them

●――――――――●

line symmetry [līn simʹə•trē] **simetría axial** A figure has line symmetry if it can be folded about a line so that its two parts match exactly.

linear unit [linʹē•ər yooʹnit] **unidad lineal** A measure of length, width, height, or distance

liquid volume [likʹwid välʹyoom] **volumen de un líquido** The amount of liquid in a container

liter (L) [lētʹər] **litro (L)** A metric unit used to measure capacity; 1 liter = 1,000 milliliters

mass [mas] **masa** The amount of matter in an object

meter (m) [mētʹər] **metro (m)** A metric unit used to measure length or distance; 1 meter = 100 centimeters

mile (mi) [mīl] **milla (mi)** A customary unit used to measure length or distance; 5,280 feet = 1 mile

milligram (mg) [milʹi•gram] **miligramo** A metric unit used to measure mass; 1,000 milligrams = 1 gram

milliliter (mL) [milʹi•lēt•ər] **mililitro (mL)** A metric unit used to measure capacity; 1,000 milliliters = 1 liter

millimeter (mm) [milʹi•mēt•ər] **milímetro (mm)** A metric unit used to measure length or distance; 1,000 millimeters = 1 meter

million [milʹyən] **millón** 1,000 thousands; written as 1,000,000

mixed number [mikst numʹbər] **número mixto** A number that is made up of a whole number and a fraction
Example: $1\frac{5}{8}$

multiple [mulʹtə•pəl] **múltiplo** The product of two counting numbers is a multiple of each of those numbers

multiplication [mul•tə•pli•kāʹshən] **multiplicación** A process to find the total number of items made up of equal-sized groups, or to find the total number of items in a given number of groups. It is the inverse operation of division.

multiply [mulʹtə•plī] **multiplicar** When you combine equal groups, you can multiply to find how many in all; the inverse operation of division

nonagon [nänʹə•gän] **eneágono** A polygon with nine sides and nine angles

not equal to (≠) [not ēʹkwəl too] **no igual a** A symbol that indicates one quantity is not equal to another

number line [numʹbər līn] **recta numérica** A line on which numbers can be located
Example:

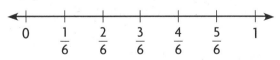

numerator [nōō′mər·āt·ər] **numerador** The number above the bar in a fraction that tells how many equal parts of the whole or group are being considered

Example:

numerical expression [nōō·mer′i·kəl ek·spresh′ən] **expresión numérica** A mathematical phrase that uses only numbers and operation signs

O

obtuse angle [äb·tōōs′ ang′gəl] **ángulo obtuso** An angle whose measure is greater than 90° and less than 180°
Example:

obtuse triangle [äb·tōōs′ trī′ang·gəl] **triángulo obtusángulo** A triangle that has one obtuse angle

octagon [äk′tə·gän] **octágono** A polygon with eight sides and eight angles
Examples:

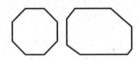

octagonal prism [äk·tag′ə·nəl priz′əm] **prisma octagonal** A three-dimensional figure with two octagonal bases and eight rectangular faces

odd [od] **impar** A whole number that has a 1, 3, 5, 7, or 9 in the ones place

open figure [ō′pən fig′yər] **figura abierta** A figure that does not begin and end at the same point

order of operations [ôr′dər əv äp·ə·rā′shənz] **orden de las operaciones** A special set of rules which gives the order in which calculations are done in an expression

ordered pair [ôr′dərd pâr] **par ordenado** A pair of numbers used to locate a point on a grid. The first number tells the left-right position and the second number tells the up-down position

origin [ôr′ə·jin] **origen** The point where the two axes of a coordinate grid intersect; (0, 0)

ounce (oz) [ouns] **onza (oz)** A customary unit used to measure weight; 16 ounces = 1 pound

overestimate [ō′vər·es·tə·mit] **sobrestimar** An estimate that is greater than the exact answer

P

pan balance [pan bal′əns] **balanza de platillos** An instrument used to weigh objects and to compare the weights of objects

parallel lines [pâr′ə·lel līnz] **líneas paralelas** Lines in the same plane that never intersect and are always the same distance apart
Example:

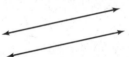

parallelogram [pâr·ə·lel′ə·gram] **paralelogramo** A quadrilateral whose opposite sides are parallel and have the same length, or are congruent
Example:

parentheses [pə·ren′thə·sēz] **paréntesis** The symbols used to show which operation or operations in an expression should be done first

partial product [pär′shəl präd′əkt] **producto parcial** A method of multiplying in which the ones, tens, hundreds, and so on are multiplied separately and then the products are added together

partial quotient [pär′shəl kwō′shənt] **cociente parcial** A method of dividing in which multiples of the divisor are subtracted from the dividend and then the quotients are added together

pattern [pat′ərn] **patrón** An ordered set of numbers or objects; the order helps you predict what will come next
Examples: 2, 4, 6, 8, 10

pentagon [pen′tə•gän] **pentágono** A polygon with five sides and five angles
Examples:

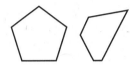

pentagonal prism [pen•tag′ə•nəl priz′əm] **prisma pentagonal** A three-dimensional figure with two pentagonal bases and five rectangular faces

pentagonal pyramid [pen•tag′ə•nəl pir′ə•mid] **pirámide pentagonal** A pyramid with a pentagonal base and five triangular faces

perimeter [pə•rim′ə•tər] **perímetro** The distance around a closed plane figure

period [pir′ē•əd] **período** Each group of three digits separated by commas in a multi-digit number
Example: 85,643,900 has three periods.

perpendicular lines [pər•pən•dik′yōō•lər līnz] **líneas perpendiculares** Two lines that intersect to form four right angles
Example:

picture graph [pik′chər graf] **gráfica con dibujos** A graph that displays countable data with symbols or pictures
Example:

HOW WE GET TO SCHOOL	
Walk	✹ ✹ ✹
Ride a Bike	✹ ✹ ✹ ✹
Ride a Bus	✹ ✹ ✹ ✹ ✹ ✦
Ride in a Car	✹ ✹

Key: Each ✹ = 10 students.

pint (pt) [pīnt] **pinta** A customary unit used to measure capacity; 2 cups = 1 pint

place value [plās val′yōō] **valor posicional** The value of each digit in a number based on the location of the digit

plane [plān] **plano** A flat surface that extends without end in all directions
Example:

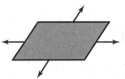

plane figure [plān fig′yər] **figura plana** See *two-dimensional figure*

point [point] **punto** An exact location in space

polygon [päl′i•gän] **polígono** A closed plane figure formed by three or more line segments
Examples:

Polygons   Not Polygons

polyhedron [päl•i•hē′drən] **poliedro** A solid figure with faces that are polygons
Examples:

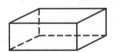

pound (lb) [pound] **libra (lb)** A customary unit used to measure weight; 1 pound = 16 ounces

prime number [prīm num′bər] **número primo** A number that has exactly two factors: 1 and itself
Examples: 2, 3, 5, 7, 11, 13, 17, and 19 are prime numbers. 1 is not a prime number.

prism [priz′əm] **prisma** A solid figure that has two congruent, polygon-shaped bases, and other faces that are all rectangles
Examples:

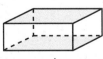

rectangular prism   triangular prism

Student Handbook H9

product [präd′əkt] **producto** The answer to a multiplication problem

protractor [prō′trak•tər] **transportador** A tool used for measuring or drawing angles

pyramid [pir′ə•mid] **pirámide** A solid figure with a polygon base and all other faces are triangles that meet at a common vertex
Example:

> **Word History**
>
> A fire is sometimes in the shape of a pyramid, with a point at the top and a wider base. This may be how *pyramid* got its name. The Greek word for fire was *pura*, which may have been combined with the Egyptian word for pyramid, *pimar*.

Q

quadrilateral [kwä•dri•lat′ər•əl] **cuadrilátero** A polygon with four sides and four angles
Example:

quart (qt) [kwôrt] **cuarto (ct)** A customary unit used to measure capacity; 2 pints = 1 quart

quotient [kwō′shənt] **cociente** The number that results from dividing
Example: 8 ÷ 4 = 2. The quotient is 2.

R

range [rānj] **rango** The difference between the greatest and least numbers in a data set

ray [rā] **semirrecta** A part of a line; it has one endpoint and continues without end in one direction
Example:

rectangle [rek′tang•gəl] **rectángulo** A parallelogram with four right angles
Example:

rectangular prism [rek•tang′gyə•lər priz′əm] **prisma rectangular** A three-dimensional figure in which all six faces are rectangles
Example:

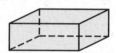

rectangular pyramid [rek•tang′gyə•lər pir′ə•mid] **pirámide rectangular** A pyramid with a rectangular base and four triangular faces

regroup [rē•grōop′] **reagrupar** To exchange amounts of equal value to rename a number
Example: 5 + 8 = 13 ones or 1 ten 3 ones

regular polygon [reg′yə•lər päl′i•gän] **polígono regular** A polygon in which all sides are congruent and all angles are congruent

related facts [ri•lāt′id fakts] **operaciones relacionadas** A set of related addition and subtraction, or multiplication and division, number sentences
Examples: 4 × 7 = 28   28 ÷ 4 = 7
7 × 4 = 28   28 ÷ 7 = 4

remainder [ri•mān′dər] **residuo** The amount left over when a number cannot be divided equally

rhombus [räm′bəs] **rombo** A parallelogram with four equal, or congruent, sides
Example:

> **Word History**
>
> *Rhombus* is almost identical to its Greek origin, *rhombos*. The original meaning was "spinning top" or "magic wheel," which is easy to imagine when you look at a rhombus, an equilateral parallelogram.

right angle [rīt ang′gəl] **ángulo recto** An angle that forms a square corner and has a measure of 90°
Example:

right triangle [rīt trī′ang•gəl] **triángulo rectángulo** A triangle that has a right angle
Example:

round [round] **redondear** To replace a number with one that is simpler and is approximately the same size as the original number
Example: 114.6 rounded to the nearest ten is 110 and to the nearest one is 115.

scale [skāl] **escala** A series of numbers placed at fixed distances on a graph to help label the graph

scalene triangle [skā′lēn trī′ang•gəl] **triángulo escaleno** A triangle with no congruent sides
Example:

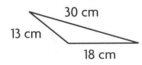

second (sec) [sek′ənd] **segundo (seg)** A small unit of time; 60 seconds = 1 minute

sequence [sē′kwəns] **sucesión** An ordered list of numbers

simplest form [sim′pləst fôrm] **mínima expresión** A fraction is in simplest form when the numerator and denominator have only 1 as a common factor.

skip count [skip kount] **contar salteado** A pattern of counting forward or backward
Example: 5, 10, 15, 20, 25, 30, . . .

solid figure [sä′lid fig′yər] **cuerpo geométrico** See *three-dimensional figure*

solution [sə•lōō′shən] **solución** A value that, when substituted for the variable, makes an equation true

sphere [sfir] **esfera** A solid figure whose curved surface is the same distance from the center to all its points
Example:

square [skwâr] **cuadrado** A polygon with four equal, or congruent, sides and four right angles

square pyramid [skwâr pir′ə•mid] **pirámide cuadrada** A solid figure with a square base and with four triangular faces that have a common vertex
Example:

square unit [skwâr yōō′nit] **unidad cuadrada** A unit used to measure area such as square foot (ft^2), square meter (m^2), and so on

standard form [stan′dərd fôrm] **forma normal** A way to write numbers by using the digits 0–9, with each digit having a place value
Example: 456 ← standard form

straight angle [strāt ang′gəl] **ángulo llano** An angle whose measure is 180°
Example:

subtraction [səb•trak′shən] **resta** The process of finding how many are left when a number of items are taken away from a group of items; the process of finding the difference when two groups are compared; the inverse operation of addition

sum [sum] **suma o total** The answer to an addition problem

Student Handbook **H11**

T

tablespoon (tbsp) [tā′bəl•spoon] **cucharada (cda)** A customary unit used to measure capacity; 3 teaspoons = 1 tablespoon

tally table [tal′ē tā′bəl] **tabla de conteo** A table that uses tally marks to record data

teaspoon (tsp) [tē′spoon] **cucharadita (cdta)** A customary unit used to measure capacity; 1 tablespoon = 3 teaspoons

tenth [tenth] **décimo** One of ten equal parts
Example: 0.7 = seven tenths

term [tûrm] **término** A number in a sequence

thousandth [thou′zəndth] **milésimo** One of one thousand equal parts
Example: 0.006 = six thousandths

three-dimensional [thrē də•men′shə•nəl] **tridimensional** Measured in three directions, such as length, width, and height

three-dimensional figure [thrē də•men′shə•nəl fig′yər] **figura tridimensional** A figure having length, width, and height
Example:

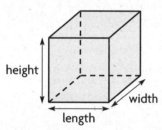

ton (T) [tun] **tonelada** A customary unit used to measure weight; 2,000 pounds = 1 ton

trapezoid [trap′i•zoid] **trapecio** A quadrilateral with at least one pair of parallel sides
Examples:

triangle [trī′ang•gəl] **triángulo** A polygon with three sides and three angles
Examples:

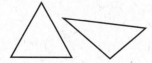

triangular prism [trī•ang′gyə•lər priz′əm] **prisma triangular** A solid figure that has two triangular bases and three rectangular faces

triangular pyramid [trī•ang′gyə•lər pir′ə•mid] **pirámide triangular** A pyramid that has a triangular base and three triangular faces

two-dimensional [too də•men′shə•nəl] **bidimensional** Measured in two directions, such as length and width

two-dimensional figure [too də•men′shə•nəl fig′yər] **figura bidimensional** A figure that lies in a plane; a figure having length and width

U

underestimate [un•dər•es′tə•mit] **subestimar** An estimate that is less than the exact answer

unit cube [yoo′nit kyoob] **cubo unitaria** A cube that has a length, width, and height of 1 unit

unit fraction [yoo′nit frak′shən] **fracción unitaria** A fraction that has 1 as a numerator

unit square [yoo′nit skwâr] **cuadrado de una unidad** A square with a side length of 1 unit, used to measure area

V

variable [vâr′ē•ə•bəl] **variable** A letter or symbol that stands for an unknown number or numbers

Venn diagram [ven dī′ə•gram] **diagrama de Venn** A diagram that shows relationships among sets of things
Example:

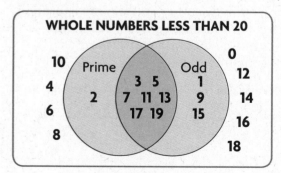

H12 Glossary

vertex [vûr′teks] **vértice** The point where two or more rays meet; the point of intersection of two sides of a polygon; the point of intersection of three (or more) edges of a solid figure; the top point of a cone; the plural of vertex is vertices
Examples:

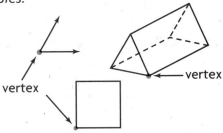

Word History

The Latin word *vertere* means "to turn" and also relates to "highest." You can turn a figure around a point, or *vertex*.

vertical [vûr′ti·kəl] **vertical** Extending up and down

volume [väl′yōōm] **volumen** The measure of the space a solid figure occupies

weight [wāt] **peso** How heavy an object is

whole [hōl] **entero** All of the parts of a shape or group

whole number [hōl num′bər] **número entero** One of the numbers 0, 1, 2, 3, 4, . . . ; the set of whole numbers goes on without end

word form [wûrd fôrm] **en palabras** A way to write numbers in standard English
Example: 4,829 = four thousand, eight hundred twenty-nine

x-axis [eks ak′sis] **eje de la *x*** The horizontal number line on a coordinate plane

x-coordinate [eks kō·ôrd′n·it] **coordenada *x*** The first number in an ordered pair; tells the distance to move right or left from (0, 0)

yard (yd) [yärd] **yarda (yd)** A customary unit used to measure length or distance; 3 feet = 1 yard

y-axis [wī ak′sis] **eje de la *y*** The vertical number line on a coordinate plane

y-coordinate [wī kō·ôrd′n·it] **coordenada *y*** The second number in an ordered pair; tells the distance to move up or down from (0, 0)

Zero Property of Multiplication [zē′rō präp′ər·tē əv mul·tə·pli·kā′shən] **propiedad del cero de la multiplicación** The property that states that when you multiply by zero, the product is zero

Student Handbook H13

Correlations

 COMMON CORE STATE STANDARDS

Standards You Will Learn

Mathematical Practices		Some examples are:
MP1	Make sense of problems and persevere in solving them.	Lessons 1.2, 1.3, 1.6, 1.7, 1.9, 2.1, 2.2, 2.3, 2.4, 2.5, 2.6, 2.8, 2.9, 3.8, 3.9, 3.11, 3.12, 4.4, 4.5, 4.7, 4.8, 5.2, 5.3, 5.4, 5.6, 6.5, 6.6, 6.7, 6.8, 6.9, 7.4, 7.6, 7.9, 7.10, 8.3, 9.1, 10.1, 10.3, 10.4, 11.1, 11.4, 11.7, 11.8, 11.10
MP2	Reason abstractly and quantitatively.	Lessons 1.1, 1.2, 1.3, 1.4, 1.5, 1.6, 1.8, 1.9, 1.11, 1.12, 2.2, 2.5, 2.6, 2.7, 3.2, 3.3, 3.6, 3.7, 3.8, 3.9, 3.12, 4.1, 4.2, 4.3, 4.6, 4.8, 5.2, 5.3, 5.4, 5.5, 5.6, 5.7, 5.8, 6.2, 6.3, 6.4, 6.5, 6.6, 6.7, 6.9, 7.3, 7.5, 7.6, 7.7, 7.8, 7.9, 8.1, 8.3, 8.5, 9.1, 10.2, 10.3, 10.5, 10.6, 11.1, 11.2, 11.5, 11.7, 11.8, 11.9
MP3	Construct viable arguments and critique the reasoning of others.	Lessons 1.3, 1.5, 1.6, 1.8, 1.9, 1.10, 1.11, 2.3, 2.4, 2.5, 2.7, 2.9, 3.4, 4.1, 4.3, 4.4, 4.6, 4.7, 5.7, 6.2, 6.3, 6.5, 7.4, 7.5, 7.6, 7.8, 7.10, 8.4, 9.3, 10.6, 11.3, 11.6, 11.11
MP4	Model with mathematics.	Lessons 1.7, 1.10, 1.11, 1.12, 2.1, 2.3, 2.7, 2.9, 3.1, 4.2, 4.5, 5.3, 5.5, 6.1, 6.2, 6.4, 6.9, 6.10, 7.2, 7.7, 9.1, 9.2, 9.3, 9.4, 9.6, 9.7, 10.2, 10.4, 10.6
MP5	Use appropriate tools strategically.	Lessons 1.1, 3.1, 3.5, 3.6, 3.7, 3.9, 3.12, 5.1, 5.2, 5.5, 5.7, 6.1, 6.2, 6.7, 7.1, 7.2, 7.3, 7.4, 7.7, 7.8, 8.1, 8.4, 8.5, 11.1, 11.5, 11.6
MP6	Attend to precision.	Lessons 1.7, 1.8, 1.10, 2.1, 2.3, 2.8, 3.3, 3.4, 3.5, 3.7, 4.1, 4.2, 4.3, 4.4, 4.8, 5.1, 5.2, 5.3, 5.4, 5.5, 5.7, 5.8, 6.1, 6.3, 6.4, 6.6, 6.7, 6.9, 7.1, 7.2, 7.3, 7.5, 7.6, 7.8, 7.9, 7.10, 8.2, 8.3, 9.2, 9.4, 9.5, 9.6, 10.1, 10.2, 10.4, 10.5, 10.7, 11.1, 11.2, 11.4, 11.5, 11.6, 11.7, 11.8, 11.9, 11.10, 11.11
MP7	Look for and make use of structure.	Lessons 1.1, 1.2, 1.4, 1.8, 2.8, 3.1, 3.2, 3.10, 5.1, 5.8, 6.8, 6.10, 9.5, 9.6, 9.7, 10.1, 10.5, 10.6, 10.7, 11.2, 11.3, 11.4, 11.10
MP8	Look for and express regularity in repeated reasoning.	Lessons 1.3, 1.4, 1.5, 2.2, 2.4, 2.6, 3.4, 3.5, 3.6, 3.8, 3.10, 4.6, 4.7, 5.6, 5.7, 6.8, 6.10, 8.2, 9.3, 9.5, 11.2, 11.4

Standards You Will Learn

Student Edition Lessons

Domain: Operations and Algebraic Thinking

Write and interpret numerical expressions.

5.OA.A.1	Use parentheses, brackets, or braces in numerical expressions, and evaluate expressions with these symbols.	Lessons 1.3, 1.10, 1.11, 1.12
5.OA.A.2	Write simple expressions that record calculations with numbers, and interpret numerical expressions without evaluating them.	Lesson 1.10, 6.4

Analyze patterns and relationships.

5.OA.B.3	Generate two numerical patterns using two given rules. Identify apparent relationships between corresponding terms. Form ordered pairs consisting of corresponding terms from the two patterns, and graph the ordered pairs on a coordinate plane.	Lessons 9.5, 9.6, 9.7

Standards You Will Learn

Student Edition Lessons

Domain: Number and Operations in Base Ten

Understand the place value system.

5.NBT.A.1	Recognize that in a multi-digit number, a digit in one place represents 10 times as much as it represents in the place to its right and 1/10 of what it represents in the place to its left.	Lessons 1.1, 1.2, 3.1
5.NBT.A.2	Explain patterns in the number of zeros of the product when multiplying a number by powers of 10, and explain patterns in the placement of the decimal point when a decimal is multiplied or divided by a power of 10. Use whole-number exponents to denote powers of 10.	Lessons 1.4, 1.5, 4.1, 5.1
5.NBT.A.3	Read, write, and compare decimals to thousandths.	
5.NBT.A.3a	Read and write decimals to thousandths using base-ten numerals, number names, and expanded form, e.g., $347.392 = 3 \times 100 + 4 \times 10 + 7 \times 1 + 3 \times (1/10) + 9 \times (1/100) + 2 \times (1/1000)$.	Lesson 3.2
5.NBT.A.3b	Compare two decimals to thousandths based on meanings of the digits in each place, using $>$, $=$, and $<$ symbols to record the results of comparisons.	Lesson 3.3
5.NBT.A.4	Use place value understanding to round decimals to any place.	Lesson 3.4

Standards You Will Learn

Student Edition Lessons

Perform operations with multi-digit whole numbers and with decimals to hundredths.		
5.NBT.B.5	Fluently multiply multi-digit whole numbers using the standard algorithm.	Lessons 1.6, 1.7
5.NBT.B.6	Find whole-number quotients of whole numbers with up to four-digit dividends and two-digit divisors, using strategies based on place value, the properties of operations, and/or the relationship between multiplication and division. Illustrate and explain the calculation by using equations, rectangular arrays, and/or area models.	Lessons 1.8, 1.9, 2.1, 2.2, 2.3, 2.4, 2.5, 2.6, 2.8, 2.9
5.NBT.B.7	Add, subtract, multiply, and divide decimals to hundredths, using concrete models or drawings and strategies based on place value, properties of operations, and/or the relationship between addition and subtraction; relate the strategy to a written method and explain the reasoning used.	Lessons 3.5, 3.6, 3.7, 3.8, 3.9, 3.10, 3.11, 3.12, 4.2, 4.3, 4.4, 4.5, 4.6, 4.7, 4.8, 5.2, 5.3, 5.4, 5.5, 5.6, 5.7, 5.8
Domain: Number and Operations—Fractions		
Use equivalent fractions as a strategy to add and subtract fractions.		
5.NF.A.1	Add and subtract fractions with unlike denominators (including mixed numbers) by replacing given fractions with equivalent fractions in such a way as to produce an equivalent sum or difference of fractions with like denominators.	Lessons 6.1, 6.4, 6.5, 6.6, 6.7, 6.8, 6.9, 6.10
5.NF.A.2	Solve word problems involving addition and subtraction of fractions referring to the same whole, including cases of unlike denominators, e.g., by using visual fraction models or equations to represent the problem. Use benchmark fractions and number sense of fractions to estimate mentally and assess the reasonableness of answers.	Lessons 6.1, 6.2, 6.3, 6.5, 6.6, 6.7, 6.9

Standards You Will Learn

Student Edition Lessons

Apply and extend previous understandings of multiplication and division to multiply and divide fractions.		
5.NF.B.3	Interpret a fraction as division of the numerator by the denominator (a/b = a ÷ b). Solve word problems involving division of whole numbers leading to answers in the form of fractions or mixed numbers, e.g., by using visual fraction models or equations to represent the problem.	Lessons 2.7, 8.3
5.NF.B.4	Apply and extend previous understandings of multiplication to multiply a fraction or whole number by a fraction.	
5.NF.B.4a	Interpret the product (a/b) × q as a parts of a partition of q into b equal parts; equivalently, as the result of a sequence of operations a × q ÷ b.	Lessons 7.1, 7.2, 7.3, 7.4, 7.6
5.NF.B.4b	Find the area of a rectangle with fractional side lengths by tiling it with unit squares of the appropriate unit fraction side lengths, and show that the area is the same as would be found by multiplying the side lengths. Multiply fractional side lengths to find areas of rectangles, and represent fraction products as rectangular areas.	Lessons 7.7, 7.10

Standards You Will Learn

Student Edition Lessons

	Apply and extend previous understandings of multiplication and division to multiply and divide fractions. *(Continued)*	
5.NF.B.5	Interpret multiplication as scaling (resizing), by:	
5.NF.B.5a	Comparing the size of a product to the size of one factor on the basis of the size of the other factor, without performing the indicated multiplication.	Lessons 7.5, 7.8
5.NF.B.5b	Explaining why multiplying a given number by a fraction greater than 1 results in a product greater than the given number (recognizing multiplication by whole numbers greater than 1 as a familiar case); explaining why multiplying a given number by a fraction less than 1 results in a product smaller than the given number; and relating the principle of fraction equivalence $a/b = (n \times a)/(n \times b)$ to the effect of multiplying a/b by 1.	Lessons 7.5, 7.6, 7.8
5.NF.B.6	Solve real world problems involving multiplication of fractions and mixed numbers, e.g., by using visual fraction models or equations to represent the problem.	Lessons 7.9, 7.10

Standards You Will Learn

Student Edition Lessons

Apply and extend previous understandings of multiplication and division to multiply and divide fractions. *(Continued)*		
5.NF.B.7	Apply and extend previous understandings of division to divide unit fractions by whole numbers and whole numbers by unit fractions.	
5.NF.B.7a	Interpret division of a unit fraction by a non-zero whole number, and compute such quotients.	Lessons 8.1, 8.5
5.NF.B.7b	Interpret division of a whole number by a unit fraction, and compute such quotients.	Lessons 8.1, 8.2, 8.5
5.NF.B.7c	Solve real world problems involving division of unit fractions by non-zero whole numbers and division of whole numbers by unit fractions, e.g., by using visual fraction models and equations to represent the problem.	Lessons 8.1, 8.4
Domain: Measurement and Data		
Convert like measurement units within a given measurement system.		
5.MD.A.1	Convert among different-sized standard measurement units within a given measurement system (e.g., convert 5 cm to 0.05 m), and use these conversions in solving multi-step, real world problems.	Lessons 10.1, 10.2, 10.3, 10.4, 10.5, 10.6, 10.7

Standards You Will Learn

Student Edition Lessons

Represent and interpret data.		
5.MD.B.2	Make a line plot to display a data set of measurements in fractions of a unit (1/2, 1/4, 1/8). Use operations on fractions for this grade to solve problems involving information presented in line plots.	Lesson 9.1
Geometric measurement: understand concepts of volume and relate volume to multiplication and to addition.		
5.MD.C.3	Recognize volume as an attribute of solid figures and understand concepts of volume measurement.	Lesson 11.4
5.MD.C.3a	A cube with side length 1 unit, called a "unit cube," is said to have "one cubic unit" of volume, and can be used to measure volume.	Lesson 11.5
5.MD.C.3b	A solid figure which can be packed without gaps or overlaps using *n* unit cubes is said to have a volume of *n* cubic units.	Lesson 11.6
5.MD.C.4	Measure volumes by counting unit cubes, using cubic cm, cubic in, cubic ft, and improvised units.	Lessons 11.6, 11.7

Standards You Will Learn

Student Edition Lessons

Geometric measurement: understand concepts of volume and relate volume to multiplication and to addition. *(Continued)*

5.MD.C.5	Relate volume to the operations of multiplication and addition and solve real world and mathematical problems involving volume.	
5.MD.C.5a	Find the volume of a right rectangular prism with whole-number side lengths by packing it with unit cubes, and show that the volume is the same as would be found by multiplying the edge lengths, equivalently by multiplying the height by the area of the base. Represent threefold whole-number products as volumes, e.g., to represent the associative property of multiplication.	Lessons 11.8, 11.9
5.MD.C.5b	Apply the formulas $V = l \times w \times h$ and $V = b \times h$ for rectangular prisms to find volumes of right rectangular prisms with whole-number edge lengths in the context of solving real world and mathematical problems.	Lessons 11.8, 11.9, 11.10
5.MD.C.5c	Recognize volume as additive. Find volumes of solid figures composed of two non-overlapping right rectangular prisms by adding the volumes of the non-overlapping parts, applying this technique to solve real world problems.	Lesson 11.11

Standards You Will Learn

Student Edition Lessons

Domain: Geometry

Graph points on the coordinate plane to solve real-world and mathematical problems.

5.G.A.1	Use a pair of perpendicular number lines, called axes, to define a coordinate system, with the intersection of the lines (the origin) arranged to coincide with the 0 on each line and a given point in the plane located by using an ordered pair of numbers, called its coordinates. Understand that the first number indicates how far to travel from the origin in the direction of one axis, and the second number indicates how far to travel in the direction of the second axis, with the convention that the names of the two axes and the coordinates correspond (e.g., x-axis and x-coordinate, y-axis and y-coordinate).	Lesson 9.2
5.G.A.2	Represent real world and mathematical problems by graphing points in the first quadrant of the coordinate plane, and interpret coordinate values of points in the context of the situation.	Lessons 9.3, 9.4

Classify two-dimensional figures into categories based on their properties.

5.G.B.3	Understand that attributes belonging to a category of two-dimensional figures also belong to all subcategories of that category.	Lessons 11.1, 11.2, 11.3
5.G.B.4	Classify two-dimensional figures in a hierarchy based on properties.	Lessons 11.1, 11.2, 11.3

Common Core State Standards © Copyright 2010. National Governors Association Center for Best Practices and Council of Chief State School Officers. All rights reserved. This product is not sponsored or endorsed by the Common Core State Standards Initiative of the National Governors Association Center for Best Practices and the Council of Chief State School Officers.

Index

A

Activities
  Activity, 23, 371, 644, 650
  Cross-Curricular. *See* Cross-Curricular Activities and Connections
  Investigate, 5, 99, 151, 175, 181, 239, 265, 297, 317, 351, 357, 427, 439, 459, 491, 545, 663, 669, 675
  Math in the Real World, 3, 85, 149, 231, 289, 349, 419, 489, 531, 583, 635
  Mental Math, 17–20, 29–32, 219, 364, 409, 598

Acute triangles, 643–646

Addition
  Associative Property of, 17–20, 219–222, 407–410
  Commutative Property of, 17–20, 219–225, 407–410
  of decimals, 175–178, 195–198, 289–210, 219–225
  estimation and, 189–192, 195–198, 363–366
  of fractions with unlike denominators, 351–354, 375–378, 407–410
  Identity Property of, 17–20
  inverse operations with subtraction, 202, 401–403
  of mixed numbers, 381–382, 395–398
  of money, 213–216
  patterns, 207–210, 395–398
  problem solving, 213–216, 401–404
  properties of, 17–20, 219–222, 407–410

Algebra
  coordinate grid
    plot ordered pairs, 539–542, 551–554, 571–574
  equations
    addition, 17–19, 351–354, 401–404
    division, 94, 517–520, 592, 623
    multiplication, 17–19, 35, 94, 517, 591, 597
    subtraction, 357–360, 401–403, 597
  expressions, 17–20, 61–64, 67–70, 73–76
  measurement
    capacity, 591–594, 611–614
    conversions, 585–588, 591–594, 597–600, 603–606, 611–614, 617–620, 623–626
    customary units, 585–588, 591–594, 597–600, 617–620
    length, 585–588, 611–614
    mass, 611–614
    metric units, 611–614, 617–620
    multistep problems, 603–606
    time, 623–626
    weight, 597–600
  patterns with decimals, 207–210, 233–236, 291–294
  volume, 669–672, 675–678, 681–684, 687–690, 693–696, 699–702

Analog clocks, 623–625

Area models, 50, 439, 459–462

Art
  Connect to Art, 448, 666

Assessment
  Chapter Review/Test, 79–84, 143–148, 225–230, 283–288, 341–346, 413–418, 483–488, 523–528, 577–582, 629–634, 705–710
  Constructed Response, 82, 148, 230, 288, 344, 418, 488, 528, 582, 634, 708
  Mid-Chapter Checkpoint, 35–36, 111–112, 187–188, 263–264, 315–316, 381–382, 457–458, 509–510, 557–558, 609–610, 661–662
  Personal Math Trainer, In every chapter. Some examples are: 3, 85, 149, 198, 248, 280, 319, 338, 354, 436, 448, 500, 514, 542, 568
  Show What You Know, 3, 85, 149, 231, 289, 349, 419, 489, 531, 583, 635

Associative Property of Addition, 17–20, 219–222, 407–410

Associative Property of Multiplication, 17–20

Average (mean), 533–536

Bar models, 137–140, 585, 591, 597, 600
Base, 23, 687–690

H24 Index

exponents, 23
prisms, 655–658, 687–690
pyramids, 655–658

Base-ten blocks, 5, 8, 23, 99–102, 175–178, 181–184, 297, 298, 309

Base-ten number system, 5–8, 11–14, 23, 99–102, 175–178, 181–184, 298

Benchmarks
to estimate decimal sums and differences, 189–192
to estimate fraction sums and differences, 363–366

Bubble maps, 290

Calculator, 220

Capacity
converting customary units, 591–594, 603–606, 617–620
converting metric units, 611–614, 617–620

Centimeters, 611–614

Chapter Openers, 3, 85, 149, 231, 289, 349, 419, 489, 531, 583, 635

Chapter Review/Test, 79–84, 143–148, 225–230, 283–288, 341–346, 413–418, 483–488, 523–528, 577–582, 629–634, 705–710

Checkpoint, Mid-Chapter. See Mid-Chapter Checkpoint

Choose a method, 219–225

Circle maps, 636

Common Core State Standards, H14–H23

Common denominators
to add and subtract fractions, 369–372, 375–378, 381–382, 383–386, 389–392, 395–398
least, 370–372

Communicate Math Ideas
Math Talk, In every lesson. Some examples are: 5, 30, 365, 604, 700
Write Math, In some lessons. Some examples are: 7, 294, 366, 593, 625, 702

Commutative Property of Addition, 17–20, 219–222, 407–410

Commutative Property of Multiplication, 17–20

Comparing
decimals, 163–166
fractions, 445–448
mixed numbers, 465–468
quadrilaterals, 649–652
two-dimensional figures, 649–652

Compatible numbers, estimate division with two-digit divisor, 113–116, 303–306

Cones, 656–658

Connect, 67, 94, 113, 131, 175, 181, 271, 277, 292, 329, 330, 375, 407, 452, 466, 503, 512, 681, 687

Connect to Art, 448, 666

Connect to Health, 32, 474

Connect to Reading, 392, 588, 658

Connect to Science, 192, 332, 554, 646

Connect to Social Studies, 102

Conversions
customary capacity, 591–594, 617–620
customary length, 585–588, 617–620
customary weight, 597–600, 617–620
metric units, 611–614, 617–620
time, 623–626

Coordinate grid
distance, 539–542
plot ordered pairs, 539–542, 545–548, 551–554, 571–574
with whole numbers, 539–542, 545–548, 551–554, 571–574

Correlations
Common Core State Standards, H14–H23

Critical Area
Common Core, 1, 347, 529

Cross-Curricular Activities and Connections
Connect to Art, 448, 666
Connect to Health, 32, 474
Connect to Reading, 392, 588, 658
Connect to Science, 192, 332, 554, 646
Connect to Social Studies, 102

Cubes, 687–690
volume of, 669–672, 675–678

Cubic units, 675–678, 687–690, 693–696

Cups, 591–594

Customary units
capacity, 585–588, 591–594, 617–620

converting, 585–588, 591–594
of length, 585–588
weight, 597–600, 617–620

Cylinders, 656–658

D

Data
collect and analyze, 533–536, 545–548, 551–554
line graphs, 551–554, 571–574
line plots, 533–536
Venn diagram, 4, 584, 638, 650

Days, 623–626

Decagonal prisms, 655

Decagons, 637–640, 655

Decimals
addition
Associative Property, 219–222
choose a method, 219–222
Commutative Property, 219–222
equivalent decimals, 196–198
estimate, 189–192, 195–198, 213–216
through hundredths, 175–178, 195–198
inverse operations, 202
model, 175–178
money, 213–216
place value and, 195–198, 219–222
regrouping, 175–178, 195–198
compare, 163–166
division
estimate, 303–306, 324, 329
model, 297–300, 309, 317–320
patterns, 291–294
place value, 309–312, 323–326
write zeros, 329–332
equivalent, 196–198
money as
addition and subtraction, 213–216
multiplication
expanded form, 251–254
model, 239–242, 245, 251–253, 265–268
money, 257–260, 278
patterns, 233–236
place value, 233–236, 245–248, 251–254, 271–274
zeros in product, 277–280
multistep problems, 335–338
order and compare, 163–166

patterns, 207–210, 233–236, 291–294
place value, 157–160, 195–198, 201–204, 219–222, 233–236, 245–248, 251–254, 271–274
rounding, 169–172, 189–192, 272
subtraction
choose a method, 219–222
equivalent decimals and, 202–204
estimate, 189–192
through hundredths, 181–184, 201–204
inverse operations, 202–204
model, 181–184
money, 213–216
place value, 201–204, 219–222
regrouping, 181–184, 201–204
thousandths
model, 151–154
read and write, 151–154, 157–160

Decimeters, 611–614, 617–620

Dekameters, 611–614

Denominators
addition, with unlike, 351–354, 375–378, 381–382, 407–410
common, 369–372, 375–378
least common denominator, 375–378, 383
subtraction, with unlike, 363–366, 375–378, 381–382

Distributive Property, 18–20, 50–51, 55–58, 252, 472–473

Division
adjusting quotients, 131–134
algorithm for, 93–96, 119–122, 309–312, 323–326, 511–514
bar models, 137–140
by decimals, 291–294, 317–320, 323–326, 329–332
of decimals, 291–294, 297–300, 303–306, 309–312, 317–320, 323–326, 329–332
Distributive Property, 50–51, 55–58
draw a diagram, 213–216, 257–260, 497–500, 517–520, 605
estimate, 87–90, 93–96, 113–116, 119–122, 131–134, 303–306, 324, 329
of four-digit numbers, 88–90, 93–112
as a fraction, 125–128, 503–506
interpreting, 517–520
interpret the remainder, 125–128
inverse operation to multiplication, 49–51, 90–95, 491–494

H26 Index

model, 49–51, 99–102, 137–140, 317–320, 491–494, 497–500, 503–506, 511–514, 517–520
by one-digit numbers, 49–51, 55–58, 87–90, 93–96, 137–140
order of operations, 67–70, 73–76
partial quotients, 105–108
patterns, 113–116, 291–294
related to multiplication, 49–51, 94, 491–494, 511–514
remainder, 94, 105–149, 125–128
of three-digit numbers, 55–58, 87–90, 93–96, 99–102, 105–108, 114–116, 119–122
by two-digit numbers, 99–102, 105–108, 113–116, 119–122, 131–134
by unit fractions, 491–494, 497–500, 511–514, 517–520
write zeros, 329–332

Divisor. *See* **Division**
one-digit divisors, 49–51, 55–58, 87–90, 93–112, 137–140
two-digit divisors, 99–102, 105–108, 113–116, 119–122, 131–134

Draw a Diagram, 213–216, 257–260, 497–500, 517–520, 605

Draw Conclusions, 6, 99, 152, 175, 181, 240, 266, 297, 317, 351, 357, 427, 439, 460, 492, 546, 670, 676

Drawing
Draw a Diagram, 213–216, 257–260, 497–500, 517–520, 605
Draw Conclusions, 6, 99, 152, 175, 181, 240, 266, 297, 317, 351, 357, 427, 439, 460, 492, 546, 670, 676

Edges
of three-dimensional figures, 663–666

Elapsed time, 623–626

Equilateral triangles, 643–646

Equivalent fractions, 369–372

Errors
What's the Error?, 14, 172, 192, 236, 300, 326, 442, 548

Essential Question, In every lesson. Some examples are: 5, 37, 351, 383, 533, 637

Estimation
decimal sums and differences, 189–192
division, two-digit divisor, 113–116
fraction sums and differences with unlike denominators, 363–366
quotients, 113–116, 303–306
volume, 681–684

Evaluate expressions, 17–20, 67–70, 73–76
with grouping symbols, 73–76

Expanded form, 11–14, 157–160, 251–254

Exponents
exponent form, 23–26
powers of 10, 23–26
word form, 23–26

Expressions
numerical, 17–20, 61–64, 67–70, 73–76
order of operations, 67–70, 73–76

Faces, 655–658, 663–666

Fahrenheit thermometer, 545–548

Feet, 585–588

Flow map, 86, 232, 490

Fluid ounces, 591–594

Formulas
for volume, 693–696, 699–702, 687–690

Fractions
addition,
Associative Property, 407–410
common denominator, 375–378, 381–382
Commutative Property, 407–410
equivalent fractions, 351–354, 375–378, 381–382, 407–410
estimate, 363–366, 381–382
mixed numbers, 381–382
models, 351–354
patterns, 395–398
properties of, 407–410
rounding, 363–366
common denominator, 369–372
as division, 503–506
draw a diagram, 497–500, 517–520
interpreting, 517–520

Index **H27**

by unit fraction, 491–494, 497–500, 511–514, 517–520
by whole-number, 491–494, 497–500, 511–514, 517–520
write an equation, 517–520
write a story problem, 518–520
equivalent, 369–372
find a fractional part of a group, 421–424
find common denominators, 369–372
least common denominator, 370–372, 375–378
line plots, 533–536
multiplication,
area, 459–462
Distributive Property, 472–473
by a fraction, 421–424, 427–430, 433–436, 439–442, 451–454
with mixed numbers, 459–462, 465–468, 471–474
models, 421–424, 427–430, 433–436, 439–442, 445–448, 451, 459–462, 465–468, 471–474
scaling, 445–448, 465–468
by whole numbers, 421–424, 427–430, 433–436
operations with line plots, 533–536
rounding, 363–366
subtraction,
common denominator, 375–378, 386–390, 389–392
equivalent fractions, 357–360, 375–378, 381–382, 389–392
estimate, 363–366
mixed numbers, 381–382, 389–392
models, 357–360
patterns, 395–398
renaming and, 389–392
rounding, 364–366
with unlike denominators, 351–354
Fraction Strips, 351–354, 357–360, 427–430, 491–494

G

Gallons, 591–594, 617–620
Geometry. *See also* Polygons; Three-dimensional figures; Two-dimensional figures
classification of figures, 649–652, 655–658
compare two-dimensional figures, 649–652
cones, 655–658
cubes, 635, 663–666, 675–678, 687–690
cylinders, 656–658
decagons, 637, 655
heptagons, 637–640
hexagons, 637–640, 655
model three-dimensional figures, 655–658, 663–666
nonagons, 637
octagons, 637–640, 655
parallelograms, 649–652
pentagons, 637–640, 655–658
polygons, 637–640, 655
polyhedrons, 655–658
prisms, 655–658, 661–664, 687–690, 693–696, 699–702
pyramids, 655–658
quadrilaterals, 637–640, 649–652
rectangles, 649–652
rhombuses, 649–652
spheres, 656–657
squares, 649–652
trapezoids, 649–652
triangles, classify, 643–646
two-dimensional figures, 637–640, 643–646, 649–652
Venn diagrams, 638, 650, 662
volume, 663–666, 669–672, 675–678, 681–684, 687–690, 693–696, 699–702

Glossary, H1–H13
Go Deeper problems, In most lessons. Some examples are: 46, 172, 320, 447, 494, 683
Grams, 611–614
Graphic Organizers. *See* Tables and Charts
Bubble Maps, 290
Circle Map, 636
Flow Map, 86, 232, 490
H-diagram, 350
problem solving, 55–56, 137–138, 213–214, 257–258, 335–336, 401–402, 477–478, 497–498, 565–566, 617–618, 693–694
Tree Map, 150, 532
Venn diagram, 4, 584, 650
Graphs,
intervals, 551–554
line graphs, 551–554, 571–574, 594
line plots, 533–536

plot ordered pairs, 539–542, 551–554, 571–574
relationships and, 571–574
Venn diagrams, 4, 584, 638, 650, 662
Grouping symbols, 73–76
Guess, check, and revise, 477–480

H-diagram, 350
Health
Connect to Health, 32, 474
Heptagons, 637–640
Hexagonal prisms, 635, 655, 657
Hexagons, 637–640, 655
Hours, 623–626

Identity Property of Addition, 17–20
Identity Property of Multiplication, 17–20, 465
Inches, 572–573, 585–588
Interpret the Remainder, 125–128
Intervals, 551–554
Inverse operations
addition and subtraction, 202, 401–402
multiplication and division, 49–51, 94–95, 491–494
Investigate, 5, 99, 151, 175, 181, 239, 265, 297, 317, 351, 357, 427, 439, 459, 491, 545, 663, 669, 675
Isosceles triangles, 643–646
***i*Tools,** 352, 359, 440

Kilograms, 611–614
Kilometers, 611–614

Lateral faces, 655–658

Least common denominator
add and subtract fractions, 375–376
finding, 370–372
Length
converting customary units, 585–588, 617–620
Lesson Essential Question, In every Student Edition lesson. Some examples are: 5, 37, 351, 383, 533, 637
Line graphs, 551–554, 571–574
Line plots, 533–536
fraction operations with, 533–536
Lines
parallel, 649
perpendicular, 649
Liters, 611–614

Make a Table, 213–216, 617–620, 693–696
Make Connections, 6, 100, 152, 176, 182, 240, 266, 298, 318, 352, 358, 428, 440, 460, 492, 546, 670, 676
Manipulatives and materials
analog clocks, 623–626
base-ten blocks, 5, 8, 23, 99–102, 175–178, 181–184, 298, 300
calculator, 220
centimeter cubes, 663, 675
Fahrenheit thermometer, 545
fraction circles, 428
fraction strips, 351–354, 357–360, 427, 491–494
MathBoard, In every lesson. Some examples are: 7, 30, 132, 390, 427, 671
number cubes, 4A, 350A, 532A
protractor, 644
ruler, 644
square tile, 459
unit cubes, 663–666, 669–672
MathBoard, In every lesson. Some examples are: 7, 30, 132, 390, 427, 671
Math in the Real World, 3, 85, 149, 231, 289, 349, 419, 489, 531, 583, 635

Index H29

Mathematical Practices
1) *Make sense of problems and persevere in solving them,* In some lessons.
Some examples are: 14, 254, 312, 536
2) *Reason abstractly and quantitatively,* In some lessons. Some examples are: 20, 386, 435, 599
3) *Construct viable arguments and critique the reasoning of others,* In some lessons. Some examples are: 40, 170, 359, 548
4) *Model with mathematics,* In some lessons. Some examples are: 90, 430, 594
5) *Use appropriate tools strategically,* In some lessons. Some examples are: 152, 494, 637
6) *Attend to precision,* In some lessons. Some examples are: 89, 474, 497, 640
7) *Look for and make use of structure,* In some lessons. Some examples are: 133, 397, 562, 646
8) *Look for and express regularity in repeated reasoning,* In some lessons. Some examples are: 170, 500, 645, 655

Math Idea, 11, 106, 637, 655

Math on the Spot Videos, In every lesson. Some examples are: 58, 154, 320, 514, 652, 684

Math Talk, In every lesson. Some examples are: 5, 30, 365, 391, 604, 700

Measurement
capacity, 591–594
conversions, 585–588, 591–594, 597–600, 603–606, 611–614, 623–626
customary units, 585–588, 591–594, 597–600
length, 29, 169, 189, 251, 271, 297, 401–402, 459–462, 477–480, 585–588, 644–645
metric units, 611–614, 617–620
multistep problems, 603–606
time, 309, 329, 623–626
volume, 675–678, 681–684, 699–702
weight, 597–600, 604–606

Mental Math, 17–20, 29–32, 219, 364, 409, 598

Meters, 611–614

Metric units
capacity, 611–614
converting, 611–614, 617–620
length, 585–588

Mid-Chapter Checkpoint, 35–36, 111–112, 187–188, 263–264, 315–316, 381–382, 457–458, 509–510, 557–558, 609–610, 661–662

Miles, 586–588

Milligrams, 611–614

Millimeters, 611–614

Millions, 11–14
place value, 11–14

Minutes, 623–626

Mixed numbers
addition of, 381–382
comparing, 465–468
multiplication of, 459–462, 465–468, 471–474
renaming as fractions, 390, 471–474
subtraction of, 381–382, 389–392

Modeling
decimal addition, 175–178
decimal division, 239–242, 245–248, 317–320
decimals, 175–178, 181–184, 317–320
Distributive Property, 18–20, 50–51
division of whole numbers, 50–51, 99–102, 137–140, 297–300, 309–312, 491–494
fraction addition, 351–354
fraction multiplication, 427–430, 433–436, 439–442, 451–454
fraction subtraction, 357–360
measurement, 585–588, 591–594, 597–600
multiplication of decimals, 239–242, 251–254, 265–268
multiplication of whole numbers, 239–242, 245, 251–254
place value, 5–8, 11–14
place value and rounding, 157–160, 169–172
three-dimensional figures, 663–666, 675–678, 681–684, 693–696
two-dimensional figures, 649–652

Modeling using bar models, 137–140, 585, 591, 597, 600

Modeling using base-ten blocks, 5–8, 23, 175–178, 181–184

Modeling using decimal models, 151–154, 175–178, 181–184, 239–242, 251–253, 265–268

H30 Index

Modeling using fraction strips, 351–354, 357–360, 427–430

Modeling using number lines, 363–366, 624–625

Modeling using place-value charts, 11–14, 157–160

Modeling using quick pictures, 100–102, 175–178, 182–184, 195–198, 240–241, 309

Modeling using Venn Diagrams, 4, 584, 650

Money
  addition and subtraction of, 213–216
  division of, 304
  estimate, 112
  multiplication of, 257–260, 277–280

Months, 623

Multiplication
  Associative Property of, 17–20, 693–696
  by decimals, 265–268, 271–274, 277–280
  decimals by whole numbers, 239–242, 245–248
    with expanded form of decimals, 251–254
  Distributive Property, 18–20, 252, 471–474
  draw a diagram, 257–260, 497–500
  estimation, 37, 272
  fraction modeling explained,
    of fractions by fractions, 439–442, 451–454
    of fractions by whole numbers, 427–430, 433–436
  as inverse operation, 49–51, 94
  inverse relationship to division, 49–51, 491–494, 511–514
  mixed numbers, 459–462, 465–468, 471–474
  models, 239–242, 251–253, 265–268
  money, 257–260, 278
  multistep problems, 603–606
  by one-digit numbers, 37–40, 491–494
  order of operations, 67–70, 73–76
  patterns, 24, 29–32, 233–236
  place value, 157–160, 233–236, 245–248, 251–254, 271–274
  by powers of 12, 23–26, 29–32, 330
  problem solving using, 55–58, 257–260, 497–500
  properties of, 17–20
  related to division, 49–51, 491–494, 511–514
  by two-digit numbers, 29–32, 43–46
  with zeros in the product, 277–280

Multistep problems,
  measurement, 603–606
  Go Deeper problems, In most lessons. Some examples are: 46, 172, 320, 447, 494, 683

Nonagons, 637

Number lines
  adding fractions on, 363
  dividing unit fractions on, 491
  estimating decimal sums and differences with, 190–192
  estimating fraction sums and differences with, 363–366
  to find elapsed time, 624
  multiplying fractions on, 446–447, 466

Numbers. *See* Decimals; Fractions; Mixed numbers; Whole numbers
  compatible, 113–116
  expanded form of, 11–14, 157–160, 251–254
  standard form of, 11–14, 157–160
  word form of, 11–14, 157–160

Number system, base ten, 5–8, 11–14, 23, 233–236, 291–294

Numerical expressions, 17–20, 61–64, 67–70
  evaluate, 67–70, 73–76

Numerical patterns, 395–398, 559–562, 571–574

Obtuse triangles, 643–646

Octagonal prisms, 655–658

Octagons, 637–640, 655

On Your Own, In most lessons. Some examples are: 13, 30, 391, 409, 639, 689

Operations
  inverse, 49–51, 94, 202, 401, 491–494, 511–514
  order of, 67–70, 73–76

Ordered pairs, 539–542, 571–574

Ordering
  decimals, 163–166

Order of Operations, 67–70, 73–76

Origin, 539–542
Ounces, 597–600, 604–606

P

Parallel lines, 649–652
Parallelograms, 649–652
Parentheses ( ), 62–64, 67–70, 73–76, 407
Partial quotients
    two-digit divisors, 105–108, 303–306
Patterns
    on a coordinate grid, 571–574
    with decimals, 207–210, 233–236, 291–294
    in division, 113–116, 291–294
    exponents, 23–26, 291
    find a rule for, 565–568, 571–574
    with fractions, 395–398
    multiplication, 24, 29–32, 233–236, 252, 271–272
    numerical, 559–562, 571–574
    place value, 5–8, 12, 246, 252, 291–294
    relate two sequences, 395–398
Pentagonal prisms, 655–658
Pentagonal pyramids, 656–658
Pentagons, 637–640
Period, 11
Perpendicular lines, 649–652
Personal Math Trainer, In every chapter. Some examples are: 3, 85, 149, 198, 248, 280, 319, 338, 354, 436, 448, 500, 514, 542, 568
Pint, 591–594, 603
Place value
    decimal addition and, 195–198, 219–225
    decimals, 157–160, 195–198, 201–204, 219–225, 233–236, 245–248, 251–254, 271–274
    decimal multiplication and, 157–160, 233–236, 245–248, 251–254, 271–274
    decimal subtraction and, 201–204, 220–222
    expanded form and, 11–14, 157–160, 251–254
    to hundred millions, 11–14
    millions, 11–14
    order decimals, 163–166
    patterns in, 5–8, 12, 233–236, 246, 252, 291–294
    round decimals, 169–172
    standard form and, 11–14, 157–160
    to thousandths, 151–154, 157–160
    whole numbers, 5–8, 11–14, 245–248, 251–254
    word form and, 11–14, 23–26, 157–160
Plane Figures, 637. *See also* Polygons; Two-dimensional figures
Polygons, 637–640, 643–646, 649–652, 655
    congruency, 638–640, 643–646, 655–658
    decagons, 637–640, 655
    heptagons, 637–640
    hexagons, 637–640, 655
    nonagons, 637
    octagons, 637–640
    pentagons, 637–640, 655–656
    quadrilaterals, 637–640, 649–652
    regular and not regular, 638–640
    triangles, 643–646, 655–656
Polyhedrons, 655–658
Pose a Problem, 96, 184, 210, 320, 430, 462, 520, 600
Pounds, 597–600, 604
Powers of 10 23–26, 29–32, 291–294
Practice and Homework
    Guided Practice, Share and Show, In every lesson. Some examples are: 7, 30, 352, 376, 645, 689
    Independent Practice, On Your Own, In most lessons. Some examples are: 13, 30, 391, 409, 639, 689
    Practice and Homework, In every lesson. Some examples are: 53–54, 179–180, 313–314, 437–438, 615–616, 703–704
    Problem Solving • Applications, In most lessons. Some examples are: 154, 192, 386, 494, 645, 666
Prerequisite skills
    Show What You Know, 3, 85, 149, 231, 289, 349, 419, 489, 531, 583, 635
Prisms, 655–658
    classifying, 655–658
    defined, 655
    rectangular, 655–658, 675–678, 681–684, 687–690, 693–696, 699–702
    triangular, 655–658
    volume of, 675–678, 681–684, 687–690, 693–696, 699–702
Problem solving
    addition and subtraction, 213–216, 401–404

customary and metric conversions, 617–620
division, 55–58, 137–140
using multiplication, 55–58, 257–260, 497–500

Problem Solving • Applications, In most lessons. Some examples are: 154, 192, 386, 494, 645, 666

Problem solving applications. *See also*
Cross-Curricular Activities and Connections
Go Deeper, In most lessons. Some examples are: 46, 172, 320, 447, 494, 683
Independent Practice. *See On Your Own*
Investigate, 5, 99, 151, 175, 181, 239, 265, 297, 317, 351, 357, 427, 439, 459, 491, 545, 663, 669, 675
Math on the Spot, In every lesson. Some examples are: 58, 154, 320, 514, 652, 684
Personal Math Trainer, In every chapter. Some examples are: 3, 85, 149, 198, 248, 280, 319, 338, 354, 436, 448, 500, 514, 542, 568
Pose a Problem, 96, 184, 210, 320, 430, 462, 520, 600
Sense or Nonsense?, 8, 20, 178, 268, 494, 574
Think Smarter, In every lesson. Some examples are: 40, 122, 306, 494, 646, 684
Think Smarter+, In every chapter. Some examples are: 177, 198, 248, 280, 319, 338, 354, 268, 436, 448, 500, 514, 542, 568
Try This!, In some lessons. Some examples are: 11, 68, 252, 364, 396, 644
Unlock the Problem, Real World, In some lessons. Some examples are: 11, 29, 363, 389, 649, 693
What's the Error?, 14, 172, 192, 236, 300, 326, 442, 548
What's the Question?, 64

Problem solving strategies
draw a diagram, 213–216, 257–260, 497–500, 517–520, 605
guess, check, and revise, 477–480
make a table, 213–216, 617–620, 693–696
solve a simpler problem, 55–58, 565–568
work backward, 335–338, 401–404

Projects, 2, 348, 530

Properties,
Associative Property of Addition, 17–20, 219, 407–410
Associative Property of Multiplication, 17–20, 693
Commutative Property of Addition, 17–20, 219, 407–410
Commutative Property of Multiplication, 17–20
Distributive Property, 18–20, 50–51, 55–58, 251–252, 472–473
Identity Property of Addition, 17–20
Identity Property of Multiplication, 17–20, 465–466

Protractors, 644

Pyramids,
classifying, 656–658
defined, 656

Quadrilaterals, 637–640, 649–652
classifying, 649–652
comparing, 649–652
defined, 649
parallelograms, 649–652
rectangles, 649–652
rhombuses, 649–652
squares, 649–652
trapezoids, 649–652

Quarts, 591–594, 603, 617, 619

Quick pictures, 23, 176–178, 182–184, 195–198, 240–241, 309

Quotients, *See Division*

Reading
Connect to Reading, 392, 588, 658
Read the Problem, 55–56, 137–138, 213–214, 257–258, 335–336, 401–402, 477–478, 497–498, 565–566, 617–618, 693–694
Visualize It, 4, 86, 150, 232, 290, 350, 420, 490, 532, 584, 636

Real World
Unlock the Problem, In most lessons. Some examples are: 11, 29, 360, 401, 593, 699

Reasonableness, 93–112, 195–198, 201–204, 239, 247, 253, 258, 363–366, 375–378, 472, 480

Rectangles
properties of, 649–652

Rectangular prisms,
properties of, 655–658
volume of, 663–666, 669–672, 675–678, 681–684, 687–690, 693–696, 699–702

Rectangular pyramids
properties of, 655–658

Regrouping
decimal addition, 175–178, 195–198, 213–216
decimal subtraction, 181–184, 201–204
division, 87–90, 93–112
multiplying, 37–40, 43–46

Regular polygons, 638–640

Relationships, mathematical
graphing, 571–574
multiplication to division, 49–51, 491–494

Remainders
in division 88–90, 93–112
interpreting, 125–128
writing as a fraction, 125–128

Remember, 12, 43, 88, 105, 189, 271, 291, 317, 364, 407

Renaming
fractions, 352, 389–392, 471
mixed numbers, 389–392, 471

Review and Test. *See* Assessment
Chapter Review/Test, 79–84, 143–148, 225–230, 389–288, 341–346, 413–418, 483–488, 523–528, 577–581, 629–634, 705–710
Mid-Chapter Checkpoint, 35–36, 111–112, 187–188, 263–264, 315–316, 381–382, 457–458, 509–510, 557–558, 609–610, 661–662
Review Words, 4, 86, 150, 232, 290, 350, 420, 490, 532, 584, 636
Show What You Know, 3, 85, 149, 231, 289, 349, 419, 489, 531, 583, 635

Rhombuses, 649–652

Right triangles, 643–646

Rounding
decimals, 169–172, 189–192, 272
fractions, 363–366
place value, 169–172

Scalene triangles, 643–646

Science
Connect to Science, 192, 332, 554, 646

Sense or Nonsense?, 8, 20, 178, 268, 494, 574

Sequences
addition, 207–210, 395–398, 559–562
pattern in, 207–210, 395–398, 559–562
relate using division, 560–561
relate using multiplication, 559–562
subtraction, 208–209, 396–398

Shapes. *See* Geometry

Share and Show, In every lesson. Some examples are: 7, 30, 352, 376, 645, 689

Show What You Know, 3, 85, 149, 231, 289, 349, 419, 489, 531, 583, 635

Simplest form
products of fractions, 433, 435–436, 452–453, 472–473
sums and differences of fractions and, 351–354, 357–360, 375–378, 408–409
sums and differences of mixed numbers and, 381–382, 389–392

Social Studies
Connect to Social Studies, 102

Solid figures. *See* Three-dimensional figures

Solve a Simpler Problem, 55–58, 565–568

Solve the Problem, 55–56, 137–138, 213–214, 257–258, 335–336, 401–402, 477–478, 497–498, 565–566, 617–618, 693–694

Spheres, 656–657

Square pyramids, 656–658

Squares
properties of, 649–652

Standard form, 11–14, 157–160

Strategies. *See* Problem solving strategies

Student help
Math Idea, 11, 106, 637, 655
Remember, 12, 43, 88, 105, 189, 271, 291, 317, 364, 407

Subtraction
of decimals, 181–184, 201–204, 208–210
estimation and, 189–192, 202–203, 363–366, 386, 389–390

of fractions with unlike denominators, 357–360, 376–378, 396–398
inverse operations with addition, 202, 401–403
of mixed numbers, 381–382, 389–392, 396–398, 401–404, 408
of money, 189–192, 213–216
patterns, 208–210, 396–398
with renaming, 389–392, 401–404

Summarize, 392

Synthesize, 175, 181

Table of Measures, H37

Tables and Charts. *See* Graphic Organizers
data, 14, 17, 20, 31, 46, 51, 64, 96, 140, 154, 160, 163, 166, 172, 192, 213–215, 222, 242, 248, 294, 306, 326, 268, 386, 424, 545, 547, 548, 551, 552, 553, 560, 561, 562, 565, 566, 567, 568, 614, 699, 700
Make a Table, 213–216, 617–620, 693–696
for measurement conversion, 586, 592, 598, 617–619, 623
place value, 5–7, 11–13, 152, 157–159, 163–165, 169–170

Talk Math. *See* Math Talk

Technology and Digital Resources
Go Digital, 4, 86, 150, 232, 350, 420, 490, 532, 584, 636
iTools, 352, 359, 440
Math on the Spot Videos, In every lesson. Some examples are: 58, 154, 320, 514, 652, 684
Multimedia eGlossary, Access through the interactive Chapter ePlanner. 4, 86, 150, 232, 290, 350, 420, 490, 532, 584, 636

Temperature, 545, 546, 551

Term, 207–210, 395–398

Test. *See* Review and Test

Think Smarter problems, In every lesson. Some examples are: 40, 122, 306, 494, 646, 684

Think Smarter+ problems, In every chapter. Some examples are: 177, 198, 248, 280, 319, 338, 354, 268, 436, 448, 500, 514, 542, 568

Thousandths
model, 151–154

read and write, 157–160

Three-dimensional figures
base, 655–658
cones, 656–657
cubes, 635, 663–666, 675–678
cylinders, 656–658
identify, describe, classify, 655–658
prisms, 655–658
pyramids, 656–658
spheres, 656–657
volume, 663–666, 669–672, 675–678, 681–684, 687–690, 693–696, 699–702

Time
elapsed, 623–626
units of, 623–626

Tons, 598–600

Trapezoids, 649–652

Tree maps, 150, 532

Triangles
acute, 643–646, 657
classifying, 643–646
equilateral, 638, 643–646
isosceles, 643–646
obtuse, 643–646
right, 643–646
scalene, 643–646

Triangular prisms, 655–658

Triangular pyramids, 656–657

Try Another Problem, 56, 138, 214, 258, 336, 402, 478, 498, 566, 618, 690, 700

Try This!, In some lessons. Some examples are: 11, 68, 252, 364, 396, 644

Two-dimensional figures
classifying, 637–640
congruency, 637–640
comparing, 649–652
polygons, 637–640
properties of, 637–640
quadrilaterals, 649–652
triangles, 643–646

Understand Vocabulary, 4, 86, 150, 232, 290, 350, 420, 490, 532, 584, 636

Unit cubes, 663–666, 675–678

Unit opener, 1–2, 347–348, 529–530

Unlike denominators
  adding, 351–354, 375–378, 381–382, 395, 397–398
  subtracting, 357–360, 376–378, 389–392, 396–398

Unlock the Problem, In most lessons. Some examples are: 11, 29, 360, 401, 649, 699

Unlock the Problem Tips, 57, 403, 695

Venn diagram, 4, 584, 650

Visualize It, 4, 86, 150, 232, 290, 350, 420, 490, 532, 584, 636

Vocabulary
  Chapter Review/Test, 79, 143, 225, 283, 341, 413, 483, 523, 577, 629, 705
  Chapter Vocabulary Cards, At the beginning of every chapter.
  Mid-Chapter Checkpoint, 35–36, 111–112, 187–188, 263–264, 315–316, 381–382, 457–458, 509–510, 557–558, 609–610, 661–662
  Multimedia eGlossary, 4, 86, 150, 232, 290, 350, 420, 490, 532, 584, 636
  Understand Vocabulary, 4, 86, 150, 232, 290, 350, 420, 490, 532, 584, 636
  Vocabulary Builder, 4, 86, 150, 232, 290, 350, 420, 490, 532, 584, 636
  Vocabulary Game, 4A, 86A, 150A, 232A, 290A, 350A, 420A, 490A, 532A, 584A, 636A
  Vocabulary Preview, 4, 150, 350, 532, 584, 636
  Vocabulary Review, 4, 86, 150, 232, 290, 350, 420, 490, 532, 584, 636

Volume
  comparison of, 675–678, 693–696
  composed figures, 699–702
  cubic unit, 675–678
  estimate, 681–684
  formula, 687–690, 693–696, 699–702, 681–684
  of rectangular prisms, 663–666, 669–672, 681–684, 687–690, 693–696, 699–702
  unit cube, 663–666

Weight
  converting customary units, 597–600, 604–606

What If, 31, 57, 68, 127, 166, 175, 181, 214–215, 246, 251, 252, 259, 271, 337, 398, 403, 478, 479, 567, 591, 603, 619, 682

What's the Error?, 14, 172, 192, 236, 300, 306, 326, 442, 548

Whole numbers
  divide decimals by, 297–300, 309–312
  divide unit fractions by, 511–514, 517–520
  divide by unit fractions, 491–494, 497–500, 517–520
  dividing, 87–90, 93–112, 99–101, 105–108, 119–122, 125–128, 131–134, 137–140
  multiply fractions by, 427–430, 433–436
  multiplying, 37–40, 43–46, 239–242, 427–430, 433–436
  place value, 5–8, 11–14, 195–198, 201–204, 245–248, 251–254
  relate multiplication to division of, 49–51, 491–494
  standard form, 11–14, 157–160
  word form of, 11–14, 157–160

Word form of numbers, 11–14, 23–25, 157–160

Work backward, 335–338, 401–404

Write Math, In every Student Edition lesson. Some examples are: 7, 269, 294, 366, 593, 621, 625, 702

Writing
  Write Math, In every Student Edition lesson. Some examples are: 7, 294, 366, 593, 625, 702

x-**axis,** 539–542, 546
x-**coordinate,** 539–542, 545–548

Yards, 585–588
y-**axis,** 539–542
y-**coordinate,** 539–542, 545–548

Table of Measures

METRIC	CUSTOMARY
Length	
1 centimeter (cm) = 10 millimeters (mm) 1 meter (m) = 1,000 millimeters 1 meter = 100 centimeters 1 meter = 10 decimeters (dm) 1 kilometer (km) = 1,000 meters	1 foot (ft) = 12 inches (in.) 1 yard (yd) = 3 feet, or 36 inches 1 mile (mi) = 1,760 yards, or 5,280 feet
Capacity	
1 liter (L) = 1,000 milliliters (mL) 1 metric cup = 250 milliliters 1 liter = 4 metric cups 1 kiloliter (kL) = 1,000 liters	1 cup (c) = 8 fluid ounces (fl oz) 1 pint (pt) = 2 cups 1 quart (qt) = 2 pints, or 4 cups 1 gallon (gal) = 4 quarts
Mass/Weight	
1 gram (g) = 1,000 milligrams (mg) 1 gram = 100 centigrams (cg) 1 kilogram (kg) = 1,000 grams	1 pound (lb) = 16 ounces (oz) 1 ton (T) = 2,000 pounds

TIME

1 minute (min) = 60 seconds (sec)
1 half hour = 30 minutes
1 hour (hr) = 60 minutes
1 day = 24 hours
1 week (wk) = 7 days
1 year (yr) = 12 months (mo), or about 52 weeks
1 year = 365 days
1 leap year = 366 days
1 decade = 10 years
1 century = 100 years
1 millennium = 1,000 years

SYMBOLS

=	is equal to	$\overleftrightarrow{AB}$	line AB
≠	is not equal to	$\overrightarrow{AB}$	ray AB
>	is greater than	$\overline{AB}$	line segment AB
<	is less than	∠ABC	angle ABC, or angle B
(2, 3)	ordered pair (x, y)	△ABC	triangle ABC
⊥	is perpendicular to	°	degree
∥	is parallel to	°C	degrees Celsius
		°F	degrees Fahrenheit

FORMULAS

	Perimeter		**Area**
Polygon	P = sum of the lengths of sides	Rectangle	$A = b \times h$, or $A = bh$
Rectangle	$P = (2 \times l) + (2 \times w)$, or $P = 2l + 2w$		
Square	$P = 4 \times s$, or $P = 4s$		

Volume

Rectangular prism $V = B \times h$, or $V = l \times w \times h$
B = area of base shape, h = height of prism